THE BRAIN BANK OF AMERICA

An Inquiry into the Politics of Science

by

PHILLIP M. BOFFEY

With an Introduction by

RALPH NADER

McGRAW-HILL BOOK COMPANY

New York   St. Louis   San Francisco

Düsseldorf   Mexico   Toronto

Book design by Elaine Gongora.

123456789BPBP798765

Library of Congress Cataloging in Publication Data

Boffey, Philip M
   The brain bank of America.

   Includes bibliographical references.
    1. Research—United States.   2. Science and
state—United States.   3. National Academy of
Sciences, Washington, D.C.   4. National Research
Council.   I. Center for Study of Responsive Law.
II. Title.
Q180.U5B56   301.31   74-23842
ISBN 0-07-006368-0

Prepared with the cooperation of the Center for Study
of Responsive Law. All author's royalties from the sale
of this book will go to the Center for Study of
Responsive Law, P.O. Box 19367, Washington, D.C.
20036, established by Ralph Nader.

The author is grateful to the following for permission to quote passages from copyrighted material:

Atlantic Naturalist for "Pesticides and the National Academy of Sciences," in the October–December 1962 issue.

Audubon, the magazine of the National Audubon Society, for "The Academy of Sciences Lays Another Thin-Shelled Egg," from the column "The Audubon View," page 103, volume 71, issue 104, July 1969 issue, copyright © 1969; and for Roland C. Clement, "Pest Control and Wildlife Relationships," page 358, volume 64, issue 6, November–December 1962 issue, copyright © 1962.

Esquire: The Magazine for Men, for Dr. James L. Goddard, "The Drug Establishment," first published in *Esquire* magazine, March 1969.

Nature magazine, for "Merit and Power," page 549, volume 230, April 30, 1971.

Nutrition Today, for Julius M. Coon, M.D., "Protecting Our Internal Environment," pages 14–16, 28–29, volume 5, number 2, Summer 1970, copyright © 1970.

Science, for H. Brooks, "Can Science Survive in the Modern Age," pages 21–30, volume 174, October 1, 1974, copyright © 1974 by the American Association for the Advancement of Science; for R. Gillette, "Lead in the: Industry Weight on Academy Panel Challenged," pages 800–802, November 19, 1971, copyright © 1971 by the American Association for the Advancement of Science; for D. S. Greenberg, "The National Academy of Sciences: Profile of an Institution (I)," pages 222 and 226, volume 156, April 14, 1967, copyright © 1967 by the American Association for the Advancement of Science; for D. S. Greenberg, "The National Academy of Sciences: Profile of an Institution (III)," page 488–493, volume 156, April 28, 1967, copyright © 1967 by the American Association for the Advancement of Science; for P. H. Abelson, "The New Physics Report," page 479, volume 177, August 11, 1972, copyright © 1972 by the American Association for the Advancement of Science; and for A. L. Hammond, "Academy Says Energy Self-Sufficiency Unlikely," page 964, volume 184, May 31, 1974, copyright © 1974 by the American Association for the Advancement of Science.

PREFACE

"Why us?" was the general reaction when the National Academy of Sciences learned that it was to be the object of an investigation by Ralph Nader's Center for Study of Responsive Law. After all, the Academy is routinely described as the most prestigious scientific body in the country; it is traditionally the benchmark of excellence by which the shortcomings of others are measured. So why go after the reputed "best" when there are demonstrated scalawags to pursue? Besides, the Academy claims that it is essentially a private organization, so why should it be subjected to public scrutiny? "If Ralph Nader's Center for Study of Responsive Law (whatever that may be) has nothing more useful to do, I suggest that it fold up its tent," commented W. H. Bradley, a member of the Academy since 1946.[1]

The attitude was characteristically arrogant, but the questions were valid. Why, indeed, should anyone want to study the Academy? The reason is simply stated: The Academy, as will become apparent to readers of this book, operates hundreds of advisory committees that influence government decisions on important public issues. Its counsel is sought in the belief that it is the most competent, most impartial adviser available. Yet the Academy operates in such secrecy that it is difficult to know whether it is acting in the public interest or in the cause of narrower vested interests.

There is little information in the public record that would enable one to make judgments about the Academy. The Academy does make most of its reports public, but the processes by which these reports are developed remain largely a mystery. No outside academic scholar has ever analyzed the workings of the Academy advisory apparatus; no Congressional committee has ever held hearings on the Academy; no mass-circulation newspapers or magazines have taken an in-depth look at the Academy; and even the specialized journals have given the Academy only passing attention.[2] This is partly because the Academy, until recently, has not been considered a very important organization on the national scene; and partly because the Academy discourages outside scrutiny. Almost all meetings of the Academy's working committees are off limits to press and public, making it impossible for an outsider to determine at first hand how well a committee performs. Similarly, most of the internal documents relating to an advisory report—including original data submitted to the committee, conflict-of-interest statements filed by committee members, expert critiques of the report by the Academy's own reviewers, occasional dissenting opinions, correspondence, minutes of meetings, and the like—are kept hidden from public view. The Academy has a rule that only those internal documents that are more than fifty years old are available for examination upon request; anything less than fifty years old is considered privileged and can be seen only with special permission, seldom granted.[3]

The Academy's clandestine ways prompted a formal protest in April 1971 from thirty-two science journalists who, upon finding themselves barred from the annual business meeting, submitted a petition complaining that "The National Academy of Sciences, though financed largely by the U.S. government and to a large extent quasi-governmental in its effects, conducts all its business meetings in private, with no opportunity for monitoring by the press or public." The journalists urged that the Academy "open its meetings, its unclassified reports, proposed resolutions and other business to press observation and coverage."[4] Similarly, the Public Interest Campaign, a nonprofit citizens' group, filed suit against the Academy in March 1974 in an effort to gain access to the deliberations and records of a committee analyzing motor vehicle emissions.[5] Such pressures have met with resistance. The Academy president, in a September 1973 memo, did urge all advisory committees to solicit "constructive contributions" from citizens' groups and individuals whose views warranted consideration, perhaps by holding public hearings or inviting them to submit written statements.[6] But that policy, which has been hailed as a major change by many Academy staffers, does little more than urge committees to seek out relevant information. At this writing, Academy committees continue to conduct almost all their business in private.

Since there is very little independent analysis available,

the public has been forced to rely primarily on the Academy's own assessments of its performance. These tend to be too superficial and too laudatory to offer much help in a critical evaluation. A history of the Academy is being prepared by an Academy-hired historian, but those who have seen early drafts say that, like most "house" histories, it will not probe very hard for imperfections. The most extensive analysis of Academy committees yet published can be found in a 1972 Academy publication, *The Science Committee*, but this, too, tends toward adulation. Prepared by a group that included two past presidents of the Academy, and written to a large extent by a former executive officer of the Academy, the report informs us that "the science committee has a long and honorable history. As a human endeavor, it has not been exempt from human foibles. But it has nonetheless, in a remarkable and inspiring way, evoked from scientists through the years wholehearted and unselfish effort to further their calling and its usefulness to mankind."[7] Thus the need for an independent evaluation of the Academy seems obvious. Before we rely too heavily on any organization, we should know a bit more about it.

Our investigation of the Academy was launched in April 1971 and continued for more than two years. The skills we applied were primarily those of the journalist. I had spent the previous 3½ years covering science policy for *Science*, the journal of the American Association for the Advancement of Science, and before that I had served for 6½ years on *The Wall Street Journal* and other publications. I devoted all of my time to the investigation for seventeen months and part of my time for fifteen months thereafter. During the summer of 1971 I was assisted, for periods ranging from six to ten weeks for each of them, by: William Freivogel, an honors graduate in political science from Stanford University and former editor-in-chief of *The Stanford Daily*; Stephen Solomon, a Phi Beta Kappa graduate in journalism from Pennsylvania State University; and Mrs. Hedvah Shuchman, who had helped compile an authoritative bibliography on *Science, Technology and Public Policy* under the direction of Lynton K. Caldwell of the University of Indiana. We were supported by a $28,000 grant from Mrs. Edgar B. Stern, a New Orleans philanthropist, to the Center for Study of Responsive Law.

The reception we got at the Academy was mixed. Initially, Philip Handler, president of the Academy, refused to admit me for a scheduled courtesy call. Handler was apparently miffed that Nader was unable to attend the meeting as originally planned and, in the words of an NAS staffer, "felt it would demean his office" if he spoke to a Nader subordinate. Subsequently, Handler made himself available for interviews which were correct and formal.

The attitudes among other Academy members ranged from hostile to cordial. Emanuel Piore, former chief scientist at IBM and treasurer of the Academy, refused to grant an interview on the grounds that the report was predestined to be a "hatchet job" (an interesting bit of prejudgment in itself). Folke Skoog, a member of the Academy since 1956, complained that our announcement of the project had "a tone of supersophistication and superficiality so unpleasant that I very much doubt that an unprejudiced and competent evaluation will be made."[8] Samuel Silver, a member of the National Academy of Engineering since 1968, asserted: "This whole business strikes me as being rather presumptuous."[9] And John R. Pierce, a Bell Laboratories executive, suggested that the Academy should launch a simultaneous investigation of us to determine our accuracy and reliability.[10]

On the other hand, Nobelist Albert Szent-Györgyi said he was "delighted" to hear about the study.[11] Olin C. Wilson, a member of the Academy since 1960, expressed admiration for Nader and "great interest" in the report.[12] Walter M. Elsasser, an Academician since 1957, said we had "a splendid opportunity to be of service to the general public."[13] And Field Winslow, a member of the executive committee of the Academy's Division of Chemistry and Chemical Technology, commented that "constructive needling from dedicated groups such as Nader's may permit Handler to reshape and invigorate the Academy and steer it toward a more useful purpose."[14]

We were granted no special access to the workings of the advisory apparatus. We were not allowed to attend committee meetings or inspect "internal" documents that would have helped an outsider assess the Academy's performance. However, within those constraints, we received substantial cooperation. The Academy's information office gave us easy access to all "public" documents—such as published reports and newspaper clippings—and obligingly photocopied many documents we wanted. We were also allowed to interview anyone connected with the Academy who would submit to questioning; only a very few turned us down. Several individuals on the Academy staff were unfailingly energetic and helpful in responding to our requests.

It is difficult for an outsider to evaluate how well a committee has performed simply by reading its final report. The report seldom makes explicit the compromises that were reached, the issues that were ducked, or the biases that may have swayed committee members. Such flaws might often be revealed in internal files. But, lacking access to those files, we were forced to rely heavily on interviews to reconstruct the factors that influenced a committee's thinking. In all, we interviewed more than five hundred Academy representatives, government officials, independent scholars, and others who have had occasion to view the Academy in action. (Any quotations in the text that are not footnoted were obtained in interviews.) We also sent out form letters soliciting comments from some 2500 individuals who appeared likely to have some knowledge about the Academy (about two hundred replied).

An evaluation of the Academy requires criteria by which to measure its performance. The Academy is typically the yardstick by which the actions of others are judged; public figures, for example, are often castigated for failure to follow Academy advice. So how do you judge the judges? The best measure we were able to find is the opinion of experts, both within the Academy and without, who can speak with authority on the issues addressed by the Academy. Thus, we compare the Academy's reports with similar reports by equally prestigious scientific groups; we reveal the critical comments made by the Academy's own internal panels of prestigious reviewers; we surface dissenting opinions that have been suppressed by the Academy; and we quote the opinions of eminent outsiders who have had occasion to take note of Academy reports. There is no perfect yardstick for measuring the Academy's performance, but we have tried to assemble the kind of evidence that would be persuasive even to most Academicians.

Although we focused primarily upon the Academy, our findings are relevant in a broader context. Ours is a society that believes in expertise, that constantly genuflects before the presumed wisdom of experts. Advisory groups and individuals are routinely called upon to counsel everyone from the President to corporation heads to the local authorities in charge of laying sewer pipe. The public tends to assume that these expert advisers dispense some sort of objective truth, the "right" answer to the problem under consideration. But such implicit trust is misplaced. There are relatively few public policy questions whose answers are purely technical. In almost all cases, an element of informed judgment is required, and what comes out strutting as "objective" wisdom is actually the subjective opinion of those who prepared the advice. Unfortunately, those expert advisers can be just as biased and pigheaded as you and me, and they can be just as foolishly wrong as we often are. If this book leaves one message with its readers, may it be a realization that expert pronouncements must never be taken on faith. They must be subjected to the same intense scrutiny and questioning as the public applies to political pronouncements. This is no trivial matter, for the problems we highlight in this book are by no means unique to the National Academy of Sciences. They exist in even more virulent form throughout thousands of established organizations, advisory groups, and institutions that affect our lives, our politics, our technology, and our destiny.

Philip M. Boffey

ACKNOWLEDGMENTS

This manuscript profited greatly from critical readings by a number of wise commentators. I particularly want to thank my colleague, Daniel S. Greenberg, for his insights and his encouragement when the going got tough. I am also indebted to Harrison Wellford, Ralph Nader, Claire Nader, and Paul Doty for commentaries on major portions of the text, to Samuel S. Epstein, Anita Johnson, and Sidney Wolfe for comments on the food protection chapter, to Connie Jo Smith for typing a lengthy, often illegible manuscript with unfailing good humor and to Kay Boffey for critical comments and editorial assistance. Finally, I am grateful to all those Academy personnel who discretely assisted me in my work while begging me not to mention their names, a request which I hereby honor.

CONTENTS

Introduction by
RALPH NADER

The mathematician-philosopher Alfred North Whitehead once wrote that "Duty arises from the power to alter the course of events." By that standard, contemporary scientists and their related institutions have an enormous degree of duty. This is particularly true of one of the most central scientific institutions, the National Academy of Sciences, which has served as an official advisor to the federal government since 1863, a position of considerable potential influence. The Academy is regularly asked to make judgments and recommendations about a wide variety of issues upon which the government must act. These include crucial matters of health and safety, such as drugs, pesticides, food additives, and radioactive waste; major technological projects, such as the supersonic transport; and military issues, such as the defoliation campaign in Vietnam. To analyze these issues, the Academy draws upon scientists and technologists from all over the country to serve on voluntary committees that prepare advisory reports for the government. Thus the Academy can be considered the preeminent forum through which individual scientists, acting as responsible citizens and rendering their best professional judgments, can have a significant impact on major technological events. But unfortunately, as Philip Boffey documents in this study, the

Academy operates under a number of constraints that impede its ability to offer critical advice and render it subordinate to major power centers in government and industry. These constraints are by no means unique to the Academy.

The vast majority of scientists and technologists work for business and the government, while most of the remainder labor at universities. Far too many are little more than chattels of their employers. These employers provide a livelihood and the status needed for promotion and recognition. Many scientists are further tied to large organizations by their need for expensive instrumentation. Thus even those lonely voices of scientific freedom whose insights and theories spring from a self-determined priority often find that their knowledge cannot be brought to society without traversing the traffic signals of large organizations.

Among such organizations, corporations probably exert the most baleful influence on technological developments and government research programs. Federal research and development budgets are significantly shaped by business and its professional cadre, except in military affairs, where the shaping pressures are shared with other national and international influences. As the Watergate power plays illustrate, there are numerous overt and subtle ways for corporate influence to insinuate itself into governmental decision-making. The research and development area is by no means exempt from such pressures. Corporate influence is exerted by direct lobbying from outside the government; by planting corporate agents within the government, as when former or future business managers spend a few years of on-the-job training in strategic places within federal agencies; and by such indirect means as the use of private "think tanks" with close ties to the corporate world to conduct policy studies for the government. Even the supposedly "pure" and "neutral" advisory committees of the National Academy of Sciences are often subject to influences emanating from the corporate world, as Boffey's study illustrates.

Corporate pressure has often been used to head off government technological initiatives that might prove threatening to traditional economic interests. In recent decades, the lukewarm response by the Department of Agriculture toward an accelerated program of research into biological and other nonchemical controls over harmful pests and the Federal Communications Commission's long stall against the spread of cable TV reflect the influence of vested economic interests within the pesticide and broadcasting industries. Even when the government launches a rare enforcement action, corporate stamina generally outlasts the federal effort. Thus in January 1969 the Justice Department sued the domestic automobile companies and their trade association, alleging a fifteen-year collusive effort to restrain the development and marketing of exhaust-control systems. The Nixon Administration inherited this case from the Johnson Administration and settled the matter nine months later in a consent decree that stands as a mockery of the legal process.

Given such corporate influence, it is not surprising that technological developments over the past generation have been strikingly biased against both workers and consumers. Since World War II, deployment of new technology has occurred in three major areas—military, computers, and producer automation. But very little technology has been made available directly to consumers for their use and control. In housing, food, surface transportation, information retrieval, communications, health, and energy, consumers have had little voice in shaping development and consequently have received little of benefit to them. Technological changes have been made in order to enhance the profits of certain producers while the cost in terms of health, safety, politics, or economics has been imposed on consumers. Witness the contrast of highly automated automobile factories extruding automobiles that change little from year to year. Or consider the absence of significant research and development of any abundant energy source (such as solar) that threatens to displace existing investments in finite underground fuels and to offer consumers a more decentralized, controllable energy source.

An innovation that profits corporations does not necessarily advance overall consumer welfare. For evidence of that, one need look no further than the recent intense criticisms made by ecologists and nutritionists of such "advances" as synthetic fibers, detergents, new food-processing techniques, and other developments. More often than not, innovations work against the consumer. Computers, diffusing at high velocity in our economy, raise serious problems of invasion of privacy, political manipulation, and further monopolization of access to information. Yet companies are not interested in giving consumers a balance of computer power. Where is the direct computer usage by consumers and citizens for effectively dealing with the marketplace and city hall?

The same concentrations of power that make for an authoritarian political economy also make for an inhumane science-technology. Over and over again, what are euphemistically called "nontechnical obstacles to innovation" frustrate both applying the known and exploring the unknown. While the most exact systems have been developed to rescue and care for wounded soldiers on the field of battle, back in the United States thousands of Americans die each year on the highways and at home for lack of a comparable technical facility. Our ambulances are little more than Cadillacs with cots. For another poignant comparison, consider the juxtaposition side by side of precise, highly machined production processes with workers tragically exposed to chemicals, gases, noise, particulate matter, radiation, and other occupational hazards which industry shows little interest in attacking. For such companies, disease—frequently long undiscovered if discovered at all—is cheaper than prevention.

There is clearly a desperate need for a consumer perspective of science-technology to cut through the mirage of ever-larger gross national products and focus on the present and future costs which are now uncounted because unacknowledged by conventional business balance sheet accounting. For any such reconstruction to be successful, those in scientific and technological roles must be able to espouse certain values, have the freedom to exert them, the resources to apply them, and the power to prevail over expected opposition:

1. *Within the corporate structure.* Suppose a chemist believes his or her company has falsified pesticide tests submitted to the Environmental Protection Agency, or a health physicist knows of a radioactive spill at a nuclear power plant that was not reported to the Atomic Energy Commission as required. What are their rights to challenge such crimes, corruption, fraud, or pollution? As we pointed out in another volume,* employed professionals are open to dismissal, demotion, and ostracism in ways wholly unlike unionized blue-collar workers whose labor-management contracts provide for grievance procedures against arbitrary corporate action. Self-censorship is further assured when the prospects for employment by other companies are considered in the context of probable blacklisting of any "troublemakers." So the exercise of professional conscience, such as is advised in those imposing codes of ethics on office walls, is difficult in corporations where the organization's dictates are so unyielding. The absence of any developed common law protecting the

---

Whistle Blowing: The Report of the Conference on Professional Responsibility. Eds. Ralph Nader, Peter J. Petkas, Kate Blackwell (Grossman Publishers, 1972).

professional's "skill right" from arbitrary impairment has led to calls for a corporate bill of rights.*

2. *Without the corporate structure as a scientist-citizen.* Suppose a biochemist took an interest in the contamination of a community's drinking water by a company that is a major customer of his or her employer. The biochemist is called into the boss's office and firmly asked to drop such citizen moonlighting because it is jeopardizing sales. Many of the same restraints operating on the scientist inside the corporation serve to keep such off-the-job citizen involvement at bay.

3. *Within the professional societies.* These groups should be the monitor and conscience of their related industries. They could critique companies or agencies that harm or defraud the public; issue *independent* health and safety standards to guide in evaluating industrial performance; provide tribunals to assess how corporations and governments treat individuals who warn the public of hazards; and use their specialized knowledge to improve public programs. But the reality is quite to the contrary. Many societies work through standards committees whose members attend on company expenses and on company missions instead of participating as independent professionals. Far from being in the vanguard of advocacy for technology assessment, these societies rarely assume consumer protection roles even after verified disclosures. Witness their laxity in allowing aerosol emissions to threaten the atmosphere's critically important ozone shield and their indifference toward contamination of drinking water through grossly inadequate purification plants. A few professional groups, such as the American Chemical Society, are troubled by their indentured status and are beginning to emerge from their corporate cocoons. At the least, they should defend their members' rights to employment against arbitrary reprisals for "blowing the whistle" on harmful practices by the employing organization.

4. *Within the government structure.* Because government bureaucracies are not accountable *outwardly* to the citizens they are serving or misserving, scientists who dissent find little or no constituency apart from an occasional reporter or member of Congress. Accountability is *upward* through the civil service ranks to the political appointee tier, which is

---

*"Make Way for the New Organization Man," by Mack Hanan, *Harvard Business Review*, July–August 1971.

penetrated routinely by special interest lobbies. There is a pressing need, buttressed by the evidence of many media, judicial, and Congressional findings, for a government-employees rights and accountability law to assure accountability to aggrieved citizens and enforceable rights for dissenting civil servants.*

5. *Within the university-college structure.* Here is where the principal independence for scientists should be found. Certainly some significant challenge to the science politics of government and business has come from the campus. The challenge to the SST, defoliation, specific arms development, carcinogens in foods, and nuclear power hazards from professors based at such universities as Harvard, MIT, and Case Western Reserve are well known. As students form more public interest research groups, with full-time legal and technical staffs, faculties will have a continuous catalyst to encourage them to contribute in policy-making forums. Yet the long arm of the funding agency, coupled with the puzzling fact that activist scientists suffer a diminution of status before their more withdrawn peers, places decisive restraints on university-based scientists. Experience at agricultural research divisions of state universities, under heavy chemical industry influence, has underscored how much courage it takes to make a statement of truth. Clinical scientific inquiry on real-life situations could assist students in understanding the importance of public-interest roles for scientists unburdened by any pressure but that of their conscience and standards as specialized citizens.

6. *The public-interest scientist.* There are only a handful of scientists working full time *on* institutions of government, business, and unions rather than *in* them. But their numbers are increasing. In Washington, D.C., the Center for Science in the Public Interest exerts this full-time scientist-as-citizen role in monitoring agencies such as the Department of Agriculture and the Food and Drug Administration. Its professionals serve as disseminators and interpreters of scientific knowledge, such as information about nutrition and effects of lead in the air; they attend trade and professional association meetings to pose issues otherwise ignored about the impact

---

*Nader, et al., *op. cit.*, Appendix D (pp. 273–286).

The Spoiled System: A Call for Civil Service Reform, by Robert Vaughn (Charterhouse, 1975).

"A Bill of Professional Rights for Employed Engineers," by Robert T. Howard, *American Engineering* (October 1966).

of products and policies on the well-being of the consumer; and they maintain a file of scientists around the country willing to dedicate some of their time to evaluate material or testify at legislative hearings. The number of full-time public-interest scientists must proliferate into the thousands in coming years if science for the betterment of mankind is to be loosed from the irons of the parochial, if not avaricious, commands of the corporate state.

In ways which are ever expanding, scientists are the first-line trustees in the management of silent violence in an industrialized society. With Medea-like intensity, the disclosures of old and new assaults on the biosphere come forward at greater and greater frequencies. To speed the end of such violence, still more scientists must add their warnings to those of informed consumers and existing practitioners of public-interest science. Then it might be possible for a new constituency of citizen-scientists to reshape or replace the captive groups which now presume to speak for the conscience of science in the garb of corporatism. This constituency, as its numbers swelled, might well provide the impetus to force reform at the National Academy, an institution whose prestigious talents are all too often subverted from working in pursuit of the public interest.

CHAPTER ONE

Emerging Importance

"For those of us who are not expert and yet must be called upon to make decisions which involve the security of our country, which involve the expenditures of hundreds of millions or billions of dollars, we must turn, in the last resort, to objective disinterested scientists who bring a strong sense of public responsibility and public obligation. So this Academy is most important."

—President John F. Kennedy, in an extemporaneous talk at the Academy, April 25, 1961.[1]

"What is the National Academy? For the American scientific community, it is, in. part, the Established Church, the House of 'Lords, the Supreme Court, and headquarters of the politics of science."

—D. S. Greenberg, *Science*, April 14, 1967.

None of the leaders of the American scientific community was prepared for the verbal attack which hit them at the December 1970 annual meeting of the American Association for the Advancement of Science. There was nothing in the program to suggest that Stewart L. Udall, former Secretary of the Interior, planned to say anything derogatory. But Udall diverged from his assigned topic and used the platform to castigate American scientists for acting like "political eunuchs" and "allowing their findings to be used as buttresses for *status quo* thinking."[2]

Udall's chief target was the National Academy of Sciences, the nation's most prestigious scientific organization. The Academy, which operates from marble headquarters facing the Lincoln Memorial in Washington, D.C., was created by an Act of Congress in 1863 to serve as an official adviser to the federal government. It also acts as an exclusive, self-perpetuating honor society, which each year elects a limited number of scientists to membership in recognition of their outstanding, original scientific work.

The Academy is widely assumed to be a source of independent, objective advice, an organization that will speak the truth on controversial issues without yielding to the pressures of political leaders or powerful special interests. But Udall challenged the validity of this assumption. He charged that the "prestige-laden" Academy functions "all too often as a virtual puppet of government," that it defines its role "in a fashion that leaves it little room to serve as an independent, critical voice," that it has become "a mere adjunct of established institutions," and that it "dutifully provides a convenient rationale for the SST lobby, the highway contractors [and] the Defense Department." What's more, he specifically criticized the Academy's president for having made "studied efforts to put down" as upstarts those scientists who have sounded the alarm on environmental problems.

The response from the Academy was outrage. In an interview with *Nature*, the British scientific journal, Philip Handler, the Academy's president, characterized the attack as "ill considered, unfounded, unnecessary, gratuitous, unwise."[3] And in a letter to *The New York Times* (never published) Handler charged that Udall had botched several environmental issues when he was Secretary of the Interior, whereas the Academy, according to Handler, had achieved "a proud record" in providing technical advice to federal agencies.[4]

There is no question that Udall's speech was long on rhetoric and short on specifics. Later he even seemed to retract a bit. He told *Nature:* "I made the speech because the scientists I have talked with share the opinions I expressed. I have not made a detailed study. I may have overstated the case. I don't know."[5] But his speech had raised important questions, and it underlined the fact that too little is known about the strengths and weaknesses of an organization whose recommendations affect the conduct of public business.

The Academy exerts substantial influence on governmental decisions through a network of hundreds of advisory committees which serve the middle levels of the federal bureaucracy. The typical mode of operation is that a federal agency which needs help with a problem will ask the Academy for advice, the Academy will reach out into the scientific community to find experts capable of analyzing the problem, these experts will convene periodically as a committee, and ultimately the committee will produce an advisory report which is submitted to the agency for its consideration in making a decision. The committees almost never perform original research; rather, they review existing information and evaluate its significance.

The importance of these committees is difficult to estimate, but the advice that they offer, or fail to offer, would appear to exert a significant cumulative impact. Virtually every citizen is affected daily by the recommendations of Academy committees. His house has probably been built in accord with minimum property standards that were based, in part, on Academy advice to the Federal Housing Administration. The medicine prescribed by his doctor is available largely because an Academy task force, which conducted a massive review of all prescription drugs on the market, told the Food and Drug Administration it was apt to be effective. The air he breathes and the water he swims in are being purified, in part, on the basis of Academy recommendations to the Environmental Protection Agency. The food he eats may contain chemicals whose safety has been endorsed by Academy committees; it also contains nutrients recommended by other Academy committees—and even the food his dog eats probably contains nutrients recommended by an Academy committee. Moreover, hundreds of decisions made by his government, ranging from such momentous issues as whether to build a supersonic transport to such trivia as whether to use chicken feathers to stuff Army mattresses, are made only after the Academy has had an opportunity to weigh in with an opinion.

The range of advisory chores handled by the Academy is staggering. To name just a few, in recent years it has: analyzed the problems of housing people from different racial, social, and economic groups in the same neighborhood; advised the Navy on long-range technical problems affecting mine warfare and undersea warfare; recommended against a plan to expand the runways at Kennedy Airport in New York; assisted in the search for a nonaddictive substitute for opium; recommended future scientific directions for the space program; predicted the likely ecological impact of a new sea-level Panama Canal; studied the genetic vulnerability of food crops to various diseases; analyzed the design and construction of highways; and assessed the hazards of enzyme-containing detergents.

However, the Academy's influence should not be exaggerated. There are dozens of crucial issues with technological components—the decision to go to the moon, the decision to build an antiballistic missile system, the proliferation and safety of nuclear reactors—in which the Academy's role has been negligible. Moreover, the Academy is not, strictly speaking, a decision maker in any of these areas. Its function is to provide expert advice to the governmental and political leaders who do make the decisions. Often the government does not seek the Academy's advice, and often it ignores the advice once proffered. Nevertheless, the trend in recent years has been to refer more and more of the tough controversial issues to the Academy for an opinion.

Even Congress, which traditionally has made little use of the Academy, has begun turning to the Academy for help on major technological issues; it does this, typically, by writing legislation which requires federal agencies to seek the Academy's advice. Thus the Clean Air Act Amendments of 1970 required government pollution officials to get the Academy's opinion on whether it was technologically feasible for the automobile industry to meet the emissions standards specified in the Act. A 1970 military authorization bill required the Pentagon to arrange for Academy investigation of the ecological and health consequences of the defoliation campaign in Vietnam. The Comprehensive Health Manpower Training Act of 1971 required that the Secretary of Health, Education, and Welfare ask the Academy to conduct a study of the cost of educating various types of health professionals. And a 1974 appropriations bill required the Environmental Protection Agency to spend $5 million for an Academy study of the scientific basis for regulating a clean environment. "Congress looks on the Academy as a competent, independent, scientific authority, perhaps the only one around," asserts an article in *Science*.[6]

Legend has it that the Academy was called into being during the Civil War by farsighted statesmen who realized they needed the help of scientists to solve the critical problems then facing the nation. The Academy's first president, Alexander Dallas Bache, propounded this inspiring version of genesis in 1864 when he praised the "resolute" Congressmen who, "with an elevated policy worthy of the great nation which they represented, took occasion to bring the scientific men around them in council on scientific matters."[7] As late as the mid-1960s, an Academy centennial volume was still promulgating this myth when it asserted that Congress called upon fifty scientists to found the Academy because "the Civil War revealed the need of our government for scientific advice."[8]

The only trouble with that thesis is that it stands history on its head. As chronicled in A. Hunter Dupree's *Science in the Federal Government*, the classic work in its field, a cabal of scientists seized the opportunity presented by the war to push through the creation of a national academy that would be a worthy counterpart of the European academies. They drafted a bill that would accomplish their aims, listed their

5

friends as incorporators, and persuaded a now-forgotten senator who had never shown any interest in science, Henry Wilson of Massachusetts, to introduce the legislation. The bill sailed through both houses on March 3, 1868, without debate in the closing hours of a lame-duck session; President Abraham Lincoln evidently signed it into law the same evening.[9] The most eminent American scientist of the day was outraged. "I do not think," Joseph Henry wrote to a colleague, "that one or two individuals have a moral right to choose for the body of scientific men in this country who shall be the members of a National Academy and then by a political ruse obtain the sanction of a law of Congress for the act."[10]

The Act of Incorporation which Wilson sneaked through Congress—often called the Academy's charter*—gave the Academy an unusual status, close to the government but not fully part of it. It granted the Academy the power to make its own rules and elect its own members, thereby rendering the Academy independent of direct supervision by the government. But it also required the Academy to serve the government by stipulating that "The Academy shall, whenever called upon by any department of the Government, investigate, examine, experiment, and report upon any subject of science or art. . . ." The "actual expense" of these studies was to be paid by the government, but beyond that the Academy was to get "no compensation whatever" for its services.[11] In recent practice, this has been interpreted by the Academy to mean that the government pays the actual cost of a study plus an overhead charge for indirect expenses. But the Academy is not entitled to make a profit on its government work and individuals who serve on Academy committees are not generally paid; they receive only their expenses.[12]

The charter has been described as "the most fundamental source of our strength" by Frederick Seitz, immediate past president of the Academy.[13] And Frank B. Jewett, a former Academy president, exclaimed in 1947 that it is "an astounding document" because "it is one of the most, if not the most, sweeping delegations of power coupled with obli-

---

*See Appendix A for full text of charter.

gation of service to the nation which the sovereign authority has ever made to a group of citizens completely outside the control of political government."[14] That's overstating the case considerably. No legal delegation of power was actually made to the Academy. But the charter established the Academy as a unique "official" scientific adviser that is not a full-fledged part of the government.

The Academy's controversial genesis caused such resentment that the institution was forced to abandon any immediate hope of playing an influential role in government affairs. Scientists who had been frozen out by the founding cabal protested vigorously; many of the original incorporators resigned; others tried to dissolve the Academy; rival organizations were formed; and attendance at Academy meetings fell sharply. Ironically, the Academy was saved by Joseph Henry—the same Henry who had protested the "political ruse" by which the Academy was founded. Apparently fearful that the prestige of American science would suffer if the Academy died, Henry, as the Academy's second president, devised a scheme to save the new organization by changing its character. He enlarged the membership so that some of the "outs" could be taken in; and he emphasized the honorary aspects of the Academy—election was to signify important contributions to scientific knowledge—while downgrading practical service to the government.[15] Today, for most of the American scientific community, this is the most important aspect of the Academy. Every year the existing members elect a limited number of outstanding scientists to their ranks. The competition is intense, and admission to the Academy is highly prized as a measure of professional success. After the April 1974 elections, the honorary NAS had 1077 members and 138 foreign associates, the reputed cream of the scientific world. The members are scattered around at various universities, industries, and government agencies. They seldom act as a cohesive unit. Most have little to do with the actual work of the organization.

And what of the advisory function? According to historian I. Bernard Cohen, the remainder of the nineteenth century saw the Academy "weak and without political influence— the Academy was not called upon to perform the tasks for which it supposedly had been established."[16] By 1913 the

Academy had made only thirty-two advisory reports, less than one per year.[17] Some tasks, including recommendations to establish such government scientific bureaus as the Geological Survey and to promote conservation of forest lands, involved important issues. But most were trivial—the selection of stone for a customs house in Chicago, a test for the purity of whiskey, preservation of paint on Army knapsacks, how best to protect the parchment of the Declaration of Independence. Small wonder that the astronomer George Ellery Hale, upon his election in 1902, found the Academy (in the words of a friend) "a small, exclusive, relatively uninfluential body which was apparently more interested in keeping young men out of its membership than in acting as a vital force in the scientific development of the United States."[18] Or, as Hale's biographer noted: "The majority of its members came to the annual meetings, listened smugly to their own speeches, ate heartily at the banquet, then returned home to forget it until the next meeting."[19]

The ambitious Hale ultimately engineered the establishment in 1916 of the vast advisory mechanism that we know today. When World War I approached, he persuaded the Academy to offer its services to President Woodrow Wilson. Wilson accepted, and Hale then quarterbacked the formation of an auxiliary organization—known as the National Research Council (NRC)—to function as a more muscular operating arm for the traditionally feeble honor society.[20]

The most important attribute of the NRC was that it greatly increased the Academy's access to advisory talent. The NRC was designed to tap the resources of the entire scientific community. It was composed of representatives from government agencies, universities, industrial firms, foundations, and scientific and engineering societies. Moreover, it ultimately developed a sizable full-time staff in Washington, D.C., which gave it a continuing presence to which the government could refer in time of need. Whenever the government needed advice on a particular problem, it could turn to the NRC, and the NRC, through its ties with the wider technical community, could reach out and find qualified specialists wherever they might happen to be working. Thus the NRC could pull together a committee of the specialists most competent to deal with a particular problem even if they did not happen to have been elected to the Academy. "The genius of the NRC concept," according to science critic Daniel S. Greenberg, "was that it would greatly enlarge the pool of talent that the Academy could call upon for advisory services, and it would do this without in any way diluting the prestige of the Academy membership. For, while the Academicians would continue to reap the honor, the NRC advisers would do most of the work."[21]

Although the NRC was supposed to serve as the principal operating arm of the Academy, relations between the two organizations were distant and strained, with occasional open bickering, for much of the next three decades. The split continued until after the Second World War, when the Academy leaders finally gained mastery over their dissident offspring. Today, the two organizations have been largely fused together. The NAS president serves as chairman of the NRC; and the NAS policy-making Council holds a majority of the seats on the governing board of the NRC.*

At the end of World War II the Academy was still, in the words of Alfred S. Romer, who was elected in 1944, "a stuffed shirt affair, taking little positive action on anything."[23] but over the next three decades, as the government's need for scientific advice grew rapidly, the Academy increased the scope of its advisory operation many times over. The Academy's annual income jumped from $3.5 million in fiscal year 1946 to $47 million in fiscal 1974.† The size of its full-time

---

*Meanwhile, the so-called "members of the NRC" have faded into unimportance. In mid-1972 there were more than three hundred such members—representatives of technical societies, liaison representatives from government agencies, delegates from the honorific Academy, and others. In theory, these NRC members serve as a link between the Academy and the technical community. But in actuality most do little more than attend an annual meeting where the Academy staff tries to brief them on current activities. One member who attended the 1971 meeting—Walter Goldschmidt, representing the American Anthropological Association—found the experience so ridiculous that he wrote a spoof about it for *Science*.[22]

---

†The Academy gets about 80 percent of its income from the federal government. In fiscal 1971, the Academy had a total income of $35.8 million, of which $28.5 million came from federal sources, $4.6 million from such nonfederal sources as industry, foundations, professional societies, and state governments, and $659,000 from endowment and trust income.

Of the $28.5 million in federal funds, about $20.3 million came through

staff increased from 186 in July 1946 to 1107 in July 1974. The number of advisory committees in simultaneous operation reached a total of 560 in 1974, employing the services of almost 9000 scientists on a part-time basis.[24] And the Academy added major new units to its organizational structure.

The first of these was the National Academy of Engineering (NAE), which was created in 1964 in response to a campaign by engineers who felt their profession was being ignored by the science-oriented Academy. The NAE honors distinguished achievement by electing worthy engineers to membership (there were 507 members after the 1974 elections). For its first decade, the NAE appointed its own committees of experts to advise the government, but in 1974 the NAE advisory committees were merged into the NRC.

The second new unit was the Institute of Medicine (IOM), which was created in 1970 in response to pressure from medical scientists who threatened to set up their own academy unless they were given greater recognition by the NAS. The IOM also serves both an honorific and an advisory func-

---

contracts for "advisory and research activities," the sort of advisory services which are the major focus of this book. Eighteen different agencies contracted for these services, as follows: Atomic Energy Commission, $4,323,601; Department of Transportation, $4,311,230; Department of Defense, $2,787,165, of which $79,126 came from the Air Force, $1,198,479 from the Army, and $1,509,560 from the Navy; Department of Health, Education, and Welfare, $2,423,902; National Science Foundation, $2,003,-324; National Aeronautics and Space Administration, $844,201; Agency for International Development, $811,529; Department of Commerce, $691,062; Department of Housing and Urban Development, $611,015; Executive Office of the President, $463,317; Veterans Administration, $358,053; Department of Agriculture, $210,856; Department of the Interior, $198,151; Environmental Protection Agency, $107,906; General Services Administration, $75,067; Department of State, $41,510; Department of Labor, $30,071; Federal Radiation Council, $2,917. Total: $20,294,875. The AEC figure of $4.3 million includes almost $4.2 million for support of the Atomic Bomb Casualty Commission, a specialized research facility administered by the Academy for the AEC in Japan. Aside from that, the AEC made relatively little use of the Academy's advisory services.

Most of the remaining $8.2 million in federal funds was for fellowships and other support of scholars. Many agencies ask the Academy to administer their fellowship programs—they simply hand the Academy a substantial sum of money, and the Academy then picks the winners and distributes the money. The dollar amount of these fellowship awards shows up on the income and expenditure sides of the Academy budget, even though the Academy is merely passing the money through from the agencies to the recipients.

The Academy's endowment has come chiefly from the major private

tion, though in a different form than the NAS and NAE. It elects members for a specified term of years, whereas the NAS and NAE elect them for life. And it requires that all members agree to work on advisory committees, whereas the NAS and NAE impose no service obligation on those honored. After the 1974 election, the IOM had 269 members and several committees at work, within the NRC structure, on problems associated with the provision of health care and medical education.*

By the 1970s, the Academy had become a major source of scientific advice for the federal government. In fiscal year 1970, more scientists served on the Academy's committees than served on advisory committees appointed directly by government agencies—namely 8328 appointments on Academy committees in that year compared with only 6424 appointments to science advisory posts with federal agen-

---

foundations. The Carnegie Corporation of New York gave the Academy $5 million in 1919, of which $1,725,000 was used to cover the cost of the Academy building and the acquisition of other property and the remainder was applied to the endowment fund. A capital drive in the late 1960s produced $5 million from the Ford Foundation, $1 million from the Rockefeller Foundation, $1 million from the Alfred P. Sloan Foundation, and $500,000 from the Commonwealth Fund. On June 30, 1971, the endowment and trust funds had a market value of $20.9 million, of which about $17 million could be applied to general purposes while the rest was targeted for specific activities at the behest of the donors.

In addition to the endowment funds, private contributions have also largely financed the construction of the Academy headquarters at 2101 Constitution Avenue, N.W., close to the Lincoln Memorial in Washington, D.C. The original façade was built with the support of the 1919 Carnegie grant, mentioned above. Two additional wings were added in the 1960s with the help of $1 million from the Equitable Life Assurance Society, $250,000 from the Ford Foundation, $240,000 from the National Science Foundation (a government agency), $100,000 from the Rockefeller Brothers Fund, $100,000 from the Rockefeller Foundation, and a host of lesser contributions. A $4.3 million auditorium and office complex was completed in 1971 with the help of major grants of $750,000 from the National Institutes of Health (a government agency), $500,000 from the Old Dominion Foundation, $400,000 from the Sloan Foundation, and $125,000 from McDonnell Aircraft. (Figures supplied by Aaron Rosenthal, Academy comptroller.)

---

*The complex of Academy organizations often describes itself by the unwieldy title of National Academy of Sciences–National Academy of Engineering–Institute of Medicine–National Research Council (NAS–NAE–IOM–NRC); but we shall use the generic term "Academy" to describe the whole or any of its parts.

cies.[25] The Academy had become, in the words of its president, Handler, "perhaps the largest 'consulting firm' in the United States."[26] And it was engaged in a remarkable range of projects at home and abroad.

The Academy conducts so many activities that the organization's annual report for fiscal year 1970 (the most recent available in 1974) requires more than 150 pages just to list brief descriptions of the major ones. The activities are generally aimed at honoring distinguished achievement, promoting the development of science, and providing advice on societal problems.

In the honorific realm, in addition to electing worthy scientists to membership, the Academy administers numerous awards for outstanding scientific achievement. It also makes certain that, when a member dies, a fellow Academician is assigned to write a eulogistic obituary in his honor. Some scientists take these honorific functions very seriously; others consider them a bore. After listening to endless speeches and award acceptances at one Academy banquet, Albert Einstein turned to his neighbor and confided: "I have just got a new theory of eternity."[27] A more recent irreverency was uttered by mathematician Stephen Smale after attending his first Academy meeting in 1971: "It was really fantastic as these days passed," he wrote, "to see how this group of America's most celebrated scientists meeting together could be so dominated by the question of just how to increase their membership and ways to remember their dead."[28] Still, honoring outstanding achievement and memorializing the departed are accepted rituals in human communities, and in these regards the Academy is no different from the Baseball Hall of Fame or Hollywood's Oscar ceremonies.

The Academy performs some of the roles traditionally carried out by scientific societies—namely, it sponsors meetings at which scientists present papers and it publishes a scientific journal, the *Proceedings of the National Academy of Sciences*. The *Proceedings* is considered one of the world's most distinguished scientific journals. Academician Britton Chance rates it "an excellent journal . . . high on your list of useful services to the scientific community at large."[29] But the journal, ironically, has suffered occasional quality con-

trol problems because of its association with the prestigious
Academy. Most other leading scientific journals will not
publish an article unless it has first been approved by expert
referees. But the *Proceedings* for many years published
virtually anything an Academician had written himself or
virtually anything he submitted which had been written by
someone else. (Most articles in the *Proceedings*, in fact, are
written by scientists who are not members of the Academy.)
The theory seemed to be that membership in the Academy
conferred an automatic right to publish or insert articles in
the institution's journal, and that the Academicians were
such a distinguished lot that it would be difficult to find
reviewers of sufficient caliber to pass judgment on their
work. But the end result, as F. Peter Woodford, former
managing editor, revealed in a 1971 report to his editorial
board, was that some "grossly incompetent work" was being
submitted. There were complaints that some Academicians
"become senile and should be restrained from publishing
poor papers for the sake of their own reputation and the
journal's."[30] In the 1970s, the editors began cracking down
on inferior submissions by requiring that articles by non-
Academicians undergo a form of refereeing, by limiting
members to submitting no more than ten papers per year,
and by actually refusing to print one or two articles submitted by members.[31] We do not mean to imply that the *Proceedings* has been crammed with incoherent ramblings; it is
unquestionably one of the giants of the world's scientific
literature. But the public tends to perceive an aura of infal-
libility about the Academy, so it is salutory to note that a
few of the Academicians may be considered senile and foolish by their colleagues.

One of the Academy's prime functions is to promote and
strengthen the national effort in science. Unlike many foreign academies, the Academy here does not generally maintain laboratories or conduct research. Instead it tries to
foster a favorable climate within which research can flour-
ish. The Academy screens fellowship applications for a number of federal and private agencies, thereby insuring that
the winners are picked primarily on the basis of merit
rather than on the basis of political favoritism at the granting agencies. It promotes agreement on standards and
nomenclature so that scientists can converse sensibly with

one another. It monitors manpower trends with an eye toward determining whether the production of scientists and engineers is keeping pace with, or lagging behind, national needs. And, perhaps most important of all, it conducts assessments of various branches of science to determine what areas of research look promising and what areas need more financial support. The Academy is rather self-consciously an "advocate" or "promoter"—some even say "lobbyist"—when it comes to protecting the interests of the scientific community. It has consistently suggested that more money should be pumped into the support of scientific research and the education of scientists. This "promotional" role seems a proper one for the nation's leading scientific organization, but it sometimes conflicts with that other major role of the Academy—the dispensing of supposedly objective advice to the government.

In the international realm, the Academy serves as a sort of State Department for American science. It operates a substantial program of technical assistance to developing countries, administers exchange programs in which scientists from this country work abroad and visitors from foreign countries come here to do their research, acts as the United States representative to international scientific organizations, and manages U.S. participation in such global scientific undertakings as the International Biological Program and the International Geophysical Year. Occasionally it even acts as a go-between in low-level diplomatic negotiations. Academy President Handler, for example, presented proposals on space cooperation to the head of the Soviet Academy in May 1970, thus catalyzing joint planning for the docking of American and Soviet satellites.[32]

The Academy occupies a peculiar niche in the array of public and private groups that offer advice on scientific matters.

It has long been acknowledged that the United States has no single science policy and no central scientific institution. Rather, it has a pluralistic system in which science policy is set by a multitude of public bodies and private organizations that operate on different levels and pursue different ends. Some of these groups are concerned with determining appropriate policies for nurturing science—how much money should be poured into the education of scientists, the building of laboratories, the support of research. Others are concerned with using science to further public objectives, such as military preparedness, or a clean environment, or a healthy society. And many—like the Academy itself—are concerned with both types of question.

At the federal level, both Congress and the executive branch play a major role in shaping and using science. They depend—particularly the executive agencies—on their own staff scientists for technical expertise. But they also seek the advice of outside experts by hiring them as consultants, appointing them to agency advisory committees, or calling them in to testify at hearings. Most of these dispensers of scientific advice are clearly either governmental or nongovernmental in nature. But the Academy's position is ambiguous.

The Academy has some attributes that suggest it is essentially a "public" agency. It operates under a Congressional charter and a presidential executive order;* it is required, for the most part, to accept any task the government asks of it; it receives most of its funding from and does most of its work for the federal government; and it is frequently assigned tasks by legislation, as when Congress requires a government agency to request the Academy's advice.

Yet the Academy claims that it is a private, independent, nongovernmental organization. It sets its own internal policies, appoints its own members and committees, and occasionally refuses to accept a government assignment—all without reference (at least in theory) to the desires of political authorities. And while the Academy does get most of its money from the government, that money funnels in via contracts and grants to perform specific jobs. The Academy gets no direct appropriation from Congress and its budget is

---

*The NRC, which was set up during World War I, was given permanent status by an executive order issued May 11, 1918. That order, plus a slightly amended version issued in 1956, are included in Appendix B. Although the original 1863 charter had implied that the Academy was to play a largely passive role, advising "whenever called upon" by the government, the phraseology of the executive order suggested, according to some Academicians, that the NRC was invited to take the initiative.

not subject to review by the White House Office of Management and Budget or by Congress. Moreover, the Academy receives some financial support from private sources, mainly foundations and industry.

Generally, the Academy tries to play it both ways. It stresses the "official" nature of its role when it wishes to enhance the authority of its pronouncements; it points to its private nature whenever journalists or Congressmen seek its internal records. The United States Government Organization Manual probably defines the Academy as well as possible, labeling it a "quasi-official" agency, along with the Federal National Mortgage Association and the American Red Cross.

Virtually everything the Academy advisory apparatus does could also be done by other organizations and agencies, but the Academy's mode of operation gives its advice a rather unique quality. On any given problem, for example, a government agency has the option of appointing its own in-house advisory committee, or of turning to the Academy. Often the people the government would appoint to its committee might be the very same people the Academy would tap if it were given the job. But the crucial distinction—one which will be explored more fully in subsequent chapters—is that the Academy's committee would, in theory at least, be somewhat more independent of the political forces running the government.

The various "think tanks" and consulting firms—such as Rand, Battelle, or the Hudson Institute—could also presumably answer most questions asked of the Academy. But their advice would almost certainly be put together by a resident staff and would be limited by the abilities of that staff, whereas the Academy is free to draw on talent wherever it might be found. Finally, the professional societies, such as the American Association for the Advancement of Science and the American Chemical Society, might also perform many of the advisory chores of the Academy. But they lack the Academy's long experience in dealing with the government and they have few financial resources to support studies of their own. In actuality, the professional societies have made scant impact on public policy.

Thus the Academy stands as a unique and preeminent

science advisor to the Federal government. But how well it performs its advisory tasks is a matter of dispute, even among the Academicians themselves, who are ultimately responsible for the proper functioning of the Academy and whose prestige imbues the operation with its aura of authority.

CHAPTER TWO

The
Honored Elite

"It's nothing to belong, but it's hell not to."

—An opinion of Academy membership.

"I have a friend who is a very successful scientist, who has a professorship at a very good school, and who has published a lot and done a lot. And the thing he wants now more than anything—he doesn't think he can get any more—he wants the honor of his career to be capped by membership in the Academy."

—Historian Nathan Reingold, in an interview, October 12, 1971.

The National Academy of Sciences is the honorary apex of the American scientific community. Only the Nobel Prize, the acknowledged pinnacle of world scientific distinction, ranks higher in prestige than election to the Academy. This is the impression that one gets from conversations with scientists; it is also the conclusion that emerges from a systematic study. Thus, a survey of the attitudes of university physicists toward ninety-eight different awards found that they ranked the Nobel Prize first, with a prestige score of 4.98 out of a possible 5.0, and Academy membership a close second, with a prestige score of 4.22. All other awards and societies were far behind.[1] Only a small minority of scientists wins election to the Academy. After the 1974 elections, there were 1077 members of the National Academy of Sciences—less than 1 percent of the more than 125,000 Americans who hold doctorates in the natural and social sciences. The Academy constitutes "the country's most distinguished scientific association," according to a group of foreign experts who surveyed American scientific institutions. "Election to the Academy is the highest honour which can be conferred on a scientist in the United States."[2]*

---

*Although most scientists are eager to get into the Academy, a handful of those elected think so little of the honor that they resign, either because they find membership unrewarding or because they end up disagreeing with the Academy leadership on issues large or small. The Academy is loathe to discuss who rejected it and why, but the roster of distinguished dropouts is known to have included such nineteenth-century luminaries as the botanist Asa Gray, the natural scientist William Barton Rogers, the mathematician Benjamin Pierce, and the naturalist Louis Agassiz. (Some of the early dropouts rejoined later.) The twentieth-century dropouts include the philosopher William James, the computer mathematician Norbert Wiener, Nobel Prize–winning physicist Richard Feynmann, and geneticists Richard C. Lewontin and A. Bruce Wallace, the last two having resigned in protest against the Academy's handling of classified contracts. George B. Field, an astronomer, refused to join when he was elected in 1972, as did Richard Levins, a population geneticist, in 1974; both cited the Academy's participation in military contracts as a reason. Getting out of

This group of eminent members is the veneer which gives the Academy its great prestige and authority. But how accomplished are the Academy members? Are they really the best in American science? The answer seems to be that, while they are unquestionably good, they are not necessarily the most outstanding in terms of scientific originality. They are the best as certified by the existing members of the Academy, but they are not necessarily the best by more objective standards. This is because a number of factors— the structure of the Academy membership, personal jealousies, behind-the-scenes politicking, and biases of some of the Academy's officers—occasionally conspire to cloud the vision of the Academy as it searches the land for distinguished scientists who merit election.

The chief criterion for election to the Academy has traditionally been excellence in original scientific research. This emphasis on original discoveries was laid down almost a century ago by Joseph Henry, the Academy's second president. "It must not be forgotten for a moment," Henry told the 1878 meeting of the Academy, "that the basis of selection is actual scientific labor in the way of original research; that is, in making positive additions to the sum of human knowledge, connected with unimpeachable moral character. It is not social position, popularity, extended authorship, or success as an instructor in science, which entitles one to membership, but actual new discoveries."[4] This dictum remains the chief guideline for elections today, though, oddly enough, the modern Academy has no explicit statement defining just what is required to merit election to the Academy. "There are no written guidelines," says Allen V. Astin, the Academy's home secretary and chief elections expert. "The unwritten criterion is that a candidate should have made important or significant original research contributions. These are the things talked about when the members get to voting."

---

the Academy is sometimes as hard as getting in. Feynmann, who is said to have become disgusted at the in-group politicking, tried to resign for more than a decade but his name was kept on the rolls and his dues were paid by the Academy leadership until finally the Academy gave in and acknowledged his departure.[3]

The emphasis on original research leads the Academy to reject some candidates who, in the eyes of the public, are distinguished achievers. James R. Killian, Jr., the first man to hold the post of presidential science advisor and former head of the Massachusetts Institute of Technology, the nation's top technical school, has never won election to the National Academy of Sciences, apparently because his skills are considered to lie in the administrative realm rather than in original research. (After the National Academy of Engineering was formed in 1964, Killian won election to that group in 1967.) Barry Commoner, the biologist–crusader who is popularly regarded as one of the scientific leaders of the environmental movement, has never been elected despite repeated nominations, apparently because the Academy membership considers his contributions to fall in the realm of public affairs rather than original research. Similarly, Jonas Salk, developer of a killed-virus polio vaccine, has never been elected to the Academy, although Albert B. Sabin, Salk's professional antagonist and developer of a rival live-virus vaccine, has been a member for more than two decades. What kept Salk out? Part of the answer is that his research was not considered particularly original. The late Thomas M. Rivers, an Academy member and head of the hospital at the Rockefeller Institute, once explained: "Now I'm not saying that Jonas wasn't a damned good man, but there have been killed vaccines before. Lots of them."[5] Moreover, Salk's vaccine was facilitated by highly original tissue culture techniques developed by three other investigators—John Enders, Thomas Weller, and Frederick Robbins—who were honored with a Nobel Prize for their contribution in 1954. Salk may also have run afoul of the bitter rivalries which split the field of microbiology over the polio question. One anonymous Academy member told Salk's biographer: "Jonas has been nominated repeatedly and has never come close to election. The factor of original work has something to do with it, but not much. Over the years, election to the Academy has become a popularity contest. The election procedures are so complicated that, if any faction in the candidate's field opposes him, he has no chance."[6]

The elections process, from beginning to end, takes a full year and has been described as "no less mysterious than that governing admission to the Druze priesthood."[7] On the surface, the procedure seems intricate, careful, and full of checks and balances aimed at guaranteeing a fair election. But on closer inspection the process turns out to be "loaded" in favor of certain kinds of candidates. The elaborate procedure is described more fully in the Academy by-laws, but for the purposes of this discussion all we need know is that (1) nominations can only be made by members of the Academy, and, while there are several pathways by which nominations can emerge, the principal role is played by the Academy's nineteen disciplinary sections—that is, the physics section is the chief nominator of physicists, the chemistry section of chemists, and so forth; (2) a crucial role in determining who gets elected is played by five class membership committees, each with jurisdiction over specified disciplinary sections, which order the nominees in a rank list from best to worst as a guide for members of the Academy in the actual voting; (3) the Council, the Academy's board of directors, has certain powers of nomination that can be used to reward or punish those whose behavior pleases or displeases the Academy leadership; and (4) the procedure is exhaustive. That is, to be elected, a candidate must first have his name proposed for nomination by an Academy member; he must be endorsed for nomination by a substantial number of other Academicians (usually two-thirds of the members of a section voting in a series of mail ballots); he must survive intense consideration by class membership committees which haggle over his qualifications in a day of face-to-face debate among committee members; he must score highly in a mail ballot of the entire Academy membership; and then he must be formally approved in a final voice vote at the Academy meeting. The whole process is carried out in great secrecy; many candidates are not even aware that they are under consideration.

How well does this elaborate, careful process work? Most Academy members believe that the Academy seldom elects anyone who doesn't deserve the honor; they see the main failures in the system as errors of omission. "There are plenty of very fine scientists who don't get elected," says Frederick Seitz, former president of the Academy, "but only a very negligible number get in who are mediocre." On the other hand, there are many Academicians who believe the Academy elects far too many nonentities. One distinguished physicist, who is a member of the Academy and was a member of the now-defunct President's Science Advisory Committee as well, contends that "It's complete nonsense to say that election to the Academy is second only to the Nobel Prize. There's a good deal of sentiment and politicking involved. Guys get elected because of their position. It really gripes me." Similarly, Roger W. Sperry, a member of the Academy since 1960, complains that at least one section of the Academy (presumably the psychology section, of which he is a member) has been taken over by a "mediocre minority" who were elected more on the basis of "popularity and political prestige" than scientific competence. "It is fast becoming a moot question as to whether it is more creditable to be listed among the Academy members or to be with the nonmembers," he says.[8]

There are few reliable yardsticks by which to judge the quality of the Academy members as compared to scientists who have not been elected. But, if the Nobel Prize is considered to be the ultimate measure of scientific creativity, then one can find instances where the Academy failed to recognize the very best scientists in the nation until after the Nobel Foundation had first honored them. In the years between 1950 and 1973 the Nobel Prize was awarded to twelve American scientists who were not members of the Academy at the time they received the award. In almost every case, the Academy rectified this oversight by later electing the Nobelist to its ranks.* But its failure to take the

---

*The names of the Nobel laureates, followed by the date they won the prize, and, in parentheses, the date they were elected to the Academy, are as follows: Frederick C. Robbins, 1954 (1972); Polykarp Kusch, 1955 (1956); Andre F. Cournand, 1956 (1958); Dickinson W. Richards, 1956 (1956); Owen Chamberlain, 1959 (1960); Donald A. Glaser, 1960 (1962); Julius Axelrod, 1970 (1971); economist Simon Kuznets, 1971 (1972); Ivar Giaever, 1973 (1974); and economist Wassily Leontief, 1973 (1974). The late Philip S. Hench, who won the Nobel Prize for medicine and physiology in 1950, never did win election to the Academy. Leon N. Cooper, who shared the Nobel Prize for physics in 1972, had not been elected as of 1974.

It should be acknowledged that the Nobel Prize is not a perfect yardstick for measuring the Academy elections. For one thing, both awards involve subjective judgment, and in any given case the Nobel judges could be in error and the Academy electors more perceptive. Moreover, the awards are

initiative in honoring these individuals suggests that the Academy's elaborate election procedures do not always succeed in discovering the most talented and productive scientists.

By another yardstick—breadth of coverage—the Academy selection process seems deficient. The Academy does not draw its honorees from the full sweep of American science. The electoral system depends heavily on the ability of the Academicians to recognize merit in other scientists. One consequence is that the Academy members tend year after year to elect their own kind—namely, scientists working at the elite universities which dominate the Academy and specializing in the disciplines already represented in the Academy. Meanwhile, scientists in other fields or from industry, government, and lesser universities are given relatively short shrift, despite efforts by the Academy to set up alternate nominating mechanisms that might help scout up less visible talent. The Academy is also a citadel of white males. After the 1974 elections, there were two blacks and sixteen women in a total membership of 1077.

The traditional explanation for these disparities is that the Academy simply mirrors the distribution of talent in the scientific community. It looks for the best scientists and, if these happen to be white males working at elite universities, so be it. That explanation has some merit, but it assumes that those elected to the Academy have performed more brilliantly than those who have not. In actuality, the only systematic study of Academy membership indicates that "whom you know" and "where you work" may be as important as "what you've done" in determining who gets elected to the Academy. A group of political scientists,

---

not necessarily based on the same criteria. Home Secretary Allen V. Astin says that the Nobels tend to be awarded for a single outstanding achievement, while Academy election tends to reflect the accomplishments of a career, though this distinction is often blurred in practice. Moreover, a Nobel Laureate's name is sometimes already under consideration in the complicated Academy nomination process at the time he wins the Nobel—thus, in any given case, it is always possible that the Academy was beginning to recognize a candidate's worth at about the same time the Nobel jury recognized his excellence. Still, the fact that the Academy has missed recognizing the accomplishments of a number of Nobelists, and has subsequently voted them into its membership, suggests that the Academy itself recognized that it had erred.

headed by Don E. Kash, who is now director of the Science and Public Policy Program at the University of Oklahoma, studied the backgrounds of the Academy membership as of August 1969.[9] At that time there were 845 members, of whom 701, the overwhelming majority, were located at American universities, while one was at a foreign university, and 143 were in government, industry, nonprofit institutions, or private employment. Kash's most striking finding was that membership in the Academy is dominated by scientists who have been educated at and employed by a handful of departments in a small number of universities—even though other evidence (cited below) suggests that top scientific talent is spread among a broader range of institutions.

Kash found that more than 70 percent of the Academy members who had been educated in the United States had earned their highest degree at one of ten elite universities, while more than 90 percent had been trained in just twenty universities.* The key factor in election seemed to be affiliation with a particular department within a university. A mere five departments—at Caltech, Harvard, Rochester, Wisconsin, and Yale—had trained 75 percent of the scientists in the Academy's genetics section who had received their highest degree at an American university. Three departments—at Harvard, Chicago, and Yale—accounted for 73 percent in anthropology; and another three—at Harvard, Chicago, and Johns Hopkins—accounted for 64 percent in microbiology. Just two departments—at Harvard and California—had trained 55 percent in zoology; another two—at M.I.T. and Caltech—trained 54 percent in engineering; and two others—at Harvard and Chicago—trained 54 percent in psychology. The pattern of domination by a handful of university departments persists whether one considers the department from which the Academicians earned

---

*The top universities, and the number of Academicians who had earned their highest degree there, were as follows: Harvard, 127; California, all campuses, 58; Chicago, 56; Columbia, 56; M.I.T., 41; Caltech, 39; Johns Hopkins, 37; Wisconsin, 35; Yale, 35; Princeton, 33; Cornell, 25; Minnesota, 18; Illinois, 16; Pennsylvania, 15; Stanford, 15; Michigan, 14; New York University, 9; Rochester, 7; Brown, 6; Ohio State, 6. The total for the top ten was 517; for the second ten, 131; and for all other American universities, 62.

their highest degree, the department in which they were employed at the time of their election to the Academy, or the department in which they were employed at the time of Kash's survey in August 1969. The greatest across-the-board strength in the Academy was shown by Harvard, which was the only university from which some members in every section of the Academy had earned their highest degree.

An Academician might counter, "Of course. What do you expect? The good scientists are at the best schools." But Kash found that the Academy membership was even more concentrated in the elite universities than seems reasonable on the basis of other measures of where scientific talent is located. For one thing, he found that the proportion of new Academy members coming from these top schools had remained virtually constant for two decades—even though there had been a substantial increase in the number of universities granting doctoral degrees in science and a substantial increase in the size of the Academy's own membership as well, two circumstances which should presumably have increased the proportion of scientists elected to the Academy from the less prestigious institutions.[9]

For another thing, Kash found that if you look at the extent to which a scientist's work is quoted in the literature (a frequently used measure of scientific significance) it turns out that the most-cited scientists come from a somewhat broader range of institutions than do the Academy members.[10] That finding suggests that there are highly productive scientists at lesser institutions who are ignored by the Academy, while their rivals from the elite schools, whose accomplishments may be less, are elected instead.

How is it that so few departments within so few universities are able to dominate the Academy? There are several conceivable explanations, but the one favored by Kash and his colleagues is that the Academy's disciplinary sections, which play the key role in nominating candidates for the Academy, serve as a self-perpetuating in-group. Kash describes the Academy sections as mirror images of the departmental structure in the universities and concludes that "being associated with an elite department in an elite university heightens the visibility of scientists to members of the Academy." As he expressed it in the cautious wording of an academic article:

> In short, while virtually everyone seems to agree that membership in the Academy is the highest honorific recognition that can be bestowed within the U.S. social system of science, the nomination and election procedures are completely in-house. The scientific community at large can neither nominate nor elect. . . . The Academy and its sections effectively constitute an in-group. Whether by conscious design or not, the nomination and election procedures have permitted, or perhaps simply resulted in, the self-perpetuation of this group as a numerical majority of the Academy's membership.[11]

Put more bluntly, the members of the Academy tend to vote in their closest colleagues—namely, those working in the same disciplines and at the same universities as the existing members. This is not the result of a conspiracy but rather of the "natural" tendency to vote for those whom one knows.

Such cliquishness may be partially responsible for the Academy's failure to accord the applied sciences equitable representation with the basic sciences. The applied fields, of which engineering, clinical medical research, and agricultural sciences are examples, seek new knowledge which can be used to achieve a particular goal, whereas the basic sciences, including such disciplines as particle physics, pure mathematics, and astronomy, seek new knowledge regardless of the purpose to which it might be applied. The distinction between the two is rather arbitrary and in practice it is often difficult to say whether a piece of work is basic or applied. When the Academy was first formed, it included a substantial number of applied scientists, and Henry's dictum that the criterion for election should be "original research" was taken to mean original work in all scientific fields, including the applied fields. But over the years the Academy gradually became an enclave of basic scientists. This happened, Home Secretary Allen Astin believes, largely because it was easier to devise yardsticks to judge the excellence of basic scientific work—one could simply review the candidate's major publications as published in the basic scientific literature. The process also undoubtedly fed on itself. As the membership became more oriented to the basic sciences, it became less competent to judge the applied fields and less aware of outstanding applied scientists. The Academy's failure to recognize the applied fields led to mini-revolts in the engineering and medical communities, which forced the Academy to set up the National Academy of Engineering and the Institute of Medicine as subordinate units. The Academy of Sciences also broadened its own base in 1971 by adding two new membership classes—medical sciences, and behavioral and social sciences—partly because of pressures for greater representation of clinical and social scientists, and partly because the Academy realized it needed broader expertise to fulfill its advisory role adequately. But as of 1972, according to Astin, "applied science is underrepresented in the Academy membership in all areas, whether physical, medical, or behavioral."

The tendency to vote for known colleagues may also distort the proportion of scientists elected to the Academy from different scientific disciplines.[12] Astronomy seems grossly overrepresented in the Academy. In the country at large, there is only one holder of a doctoral degree in astronomy for every twenty holders of a physics doctorate—674 astronomers compared to 13,531 physicists, according to 1970 figures from the National Register of Scientific and Technical Personnel.[13] Yet in the Academy, there is better than one astronomer for every three physicists—forty members of the astronomy section compared to 111 members of the physics section as of mid-1971. Some Academicians contend this is because many of the very brightest physicists have shifted to astronomy. But an equally plausible explanation is that the astronomers were simply more successful at establishing a firm beachhead in the Academy and then continued to elect their own kind. Anthropology also seems to have had considerable success in comparison with other behavioral disciplines. In the country at large, there is only one Ph.D. anthropologist engaged in research and development work for every fourteen Ph.D. psychologists so engaged—331 anthropologists compared to 4542 psychologists in 1970.[14] Yet in the Academy the anthropology section is almost as large as the psychology section—twenty-four anthropologists compared with thirty-five psychologists as of mid-1971. Again, why the disparity? Part of the explanation may be that proportionately more psychologists are engaged in applied research, but there are still seven times as many psychologists engaged in basic research as anthropologists, so that is not the whole story.*

These numerical games should not be carried too far because they ignore a number of factors, including the relative intellectual merit of a given field and its development over a long period of years. But even after all the caveats have been uttered, it seems likely that the disproportionate representation of disciplines is at least partly due to the fact that the disciplines already represented in the Academy tend to nominate and elect their own kind. "If certain scientific areas are strongly represented in the Academy," says George B. Kistiakowsky, the Academy's vice-president from 1965 to 1973, "people working in the mainstream of those areas are more likely to be elected than others. Applied mathematicians (the computer specialists and statisticians) just were not elected for a long time. And the molecular geneticists had trouble getting in because the Academy members were classical geneticists."

"Honorary bodies have time lags," explains John T. Edsall, a member of the Academy and former editor of its scientific journal. "The long-established fields keep electing people while the rising fields lag behind. I could see it happening in biochemistry. I've been a member for twenty years. For the first ten years it seemed to me that biochemistry was very underrepresented in the Academy compared to its importance in the scientific community generally. But gradually the lag was made up, and now the biochemistry section is a very large one."

Ecology was also slow to gain recognition. Although it has developed as a broad-gauged discipline in recent years, the Academy has elected relatively few scientists who consider

---

*The *National Register* gives only partial coverage of the scientific community. It relies on questionnaires sent out to lists compiled by cooperating scientific societies. Some scientists are not on these lists—and not all of those who are on the lists respond. In 1970, 64 percent of the qualified scientists on the mailing lists returned questionnaires. This partial coverage does not affect the argument advanced here, which is simply that some disciplines are overrepresented in the Academy in comparison to others. The compilers of the National Register report that there are only "slight differences" in the reporting patterns among disciplines. See *American Science Manpower 1970*, p. 243.

themselves ecologists. There is no separate ecology section, but ecologists can be nominated by one of the specialized disciplinary sections (such as zoology or applied biology), or, starting with the 1973 election, they can be nominated by a new temporary nominating group covering environmental and field biology. Why has the election of ecologists lagged? Stanley A. Cain, director of the Institute for Environmental Quality at the University of Michigan, comments:

"I was elected to the Academy in 1970 at a time when two or three ecologists went in. I believe that ecologists were recognized tardily because the field was integrative, rather than specialized, although some ecologists had developed special areas such as energy flux and mineral cycling. Also, the universities that have traditionally provided a large share of the Academy's membership had not yet developed ecology. Furthermore, ecology was seldom 'pure.' It was logically put to work in public health, agriculture, resources management, etc., and universities with such orientation were seldom contributors of Academy members. It is my personal opinion that the Academy has had difficulty meeting its public service obligations in a field such as applied ecology."[15]

The Academy's failure to honor certain disciplines often annoys the practitioners in the neglected fields. "The NAS has many of the aspects of a self-perpetuating gentleman's club and it is not representative of the scope of American science," complains D. Mark Hegsted, a Harvard nutritionist who was not a member of the Academy when he made the comment but who was subsequently elected. "Just why a few representatives of a few areas of science should sit in judgment of those areas in which they have little or no competence should be questioned."[16] However, most Academicians seem to regard the complaints of the "outs" as pure sour grapes. "There will always be a few who claim that bias kept them out of the Academy," says Academician S. A. Goudsmit. "That is natural ... a touch of paranoia is an essential characteristic of a successful research worker."[17]

As is true of many honorific bodies, the Academy membership tends to be post-middle-aged and therefore somewhat removed from the frontiers of current research. The average age at election in recent years has hovered around fifty (for the membership which Kash studied it was forty-nine).

As of July 1, 1970, the median age of the 866 members was sixty-two. There were six members in their nineties, fifty-one in their eighties, 160 in their seventies, and 276 in their sixties. Anyone in his forties is considered a "youngster" in the Academy; there were only six members in their thirties, the youngest being thirty-six.[18] On formal occasions, such as the annual meeting, the Academy seems even more aged than it really is, for many of the younger members don't attend. At the 1971 meeting, none of the members in their thirties showed up, and less than a third of those in their forties attended.[19] "The Academy is a blend of ... active scientists and engineers, and fuddy duddies," according to Academician E. N. Parker, but "the old fossils ... dominate the formal occasions because they are not occupied with more productive activities."[20]

The age of the membership has been a recurrent cause of concern to activists who believe that seniority acts as a deadening and enfeebling influence on the Academy's activities. In recent years there has been a two-pronged effort to give the Academy a more youthful constituency. An amendment was proposed to the bylaws that would require all members to be transferred to "emeritus" status at age seventy-five—but this was defeated at the 1971 annual meeting.[21] At the other end of the age spectrum, an Academy committee recommended that quotas be established to insure the election of younger scientists, but the Council rejected this proposal and merely issued a statement calling it "important that younger scientists be better represented in the membership and activities of the Academy, both to take account of many newer scientific interests and points of view, and because of the vigorous assistance these younger scientists could provide in the advisory work of the Academy and the Research Council."[22] At the 1972 annual meeting the membership rejected a motion that the president appoint a committee to study ways whereby the age structure of the Academy might be modified through the election of more young people.[23]

Kash and his colleagues described the Academy election results in terms of visibility—Academy members tend to vote for their close colleagues rather than for researchers at remote universities or in unfamiliar fields. But visibility is not the only factor which can undermine the integrity of an election that is supposedly based solely on merit.

Intense politicking also appears to be a factor—university departments occasionally vote *en bloc* in an effort to get their colleagues elected. "If one aspires to membership in the Academy, it is very helpful to have good friends who are already members," explains Ms. Annette Cronin, whose husband, James, was elected in 1970. "If they are in one's university department, so much the better. Departments like the prestige of additional members."[24] During the nominating and elections process, the communications circuits between universities are often abuzz with campaigning. Academician Dwight J. Ingle, who believes the Academy generally does a good job in picking members, nevertheless acknowledges that "members of the NAS sometimes engage in special efforts to elect this or that scientist to membership or to prevent the election of a qualified scientist who has been nominated. Consequently elections are not always entirely fair."[25] Richard C. Lewontin, who was elected to the Academy in 1968 and resigned in 1971, reports that one colleague—A. Bruce Wallace, a Cornell geneticist—was kept out of the Academy for years because of a "vendetta" conducted by a particularly powerful Academician.

The biases of some of the Academy's highest officers can also distort the outcome of an Academy election. This showed up most dramatically at the 1971 annual meeting when the Council of the Academy engineered the rejection of Lamont C. Cole, an outspoken ecologist who had made public statements on environmental issues which Academy leaders regarded as alarmist.

Cole's name emerged from the complicated preference balloting which precedes the formal election in forty-seventh place on a rank list of seventy-five names. Since the number of new members to be elected that year was fifty, he seemed almost certain to win election. But the Council, exercising certain prerogatives it enjoys during an election, recommended that Cole's name be bypassed and that number fifty-one on the list be elected instead. The members at the meeting, after a bitter debate, approved the Council's recommendation, with only five of Cole's supporters voting nay.

President Handler explained that the Council had deliberately snubbed Cole because he had made "scientifically inaccurate public statements which had served to confuse public debate."[26] The Council was apparently thinking particularly of popular articles in which Cole had speculated that the world's oxygen supply might be disastrously depleted by widespread burning of fossil fuels, the paving over of areas that once nourished oxygen-producing green plants, and the destruction by chemical pollution of the microscopic marine plants that produce most of the world's oxygen.[27] Many prominent Academicians told the study team that they voted against Cole because they considered him a marginal candidate whose credentials were too slim to justify election. But the fact remains that Cole had emerged from the original preference balloting, which is supposedly based on a judgment of scientific contributions, with enough votes to qualify for election. His real sin was that he made statements offensive to the Council. Handler has often assailed scientists for what he regards as exaggerated warnings about ecological perils—warnings which he believes produce a backlash against science and technology for supposedly causing environmental ills.[28] Moreover, Cole had certainly not endeared himself to the Academy leadership by once proclaiming: "The National Academy doesn't know enough about ecology to know how ignorant it is."[29] Cole's indiscretion was probably not exaggeration per se. Would he have been rejected if he had made exaggerated statements which pleased the leaders of the scientific community? Would he have been dropped, for example, if he had asserted, as Handler once did in pleading for Congressional support of scientific research, that the nation's scientific apparatus was "falling into shambles"[30] because of federal budget cuts—even though Handler later declared that "our scientific capabilities were never greater; our scientific productivity remains the marvel of the world ... relatively few academic laboratories known to me have yet been really seriously injured"[31]? Would he have been snubbed if he had made exaggerated claims, as Handler did on several occasions, to the effect that science deserved high praise because "1970 marked the abolition of a once devastating disease— last year, for the first time in recorded history, not a single case of smallpox was reported anywhere in the world"[32]? Handler called that supposed feat "one of man's truly great triumphs over his ever-hostile environment" and suggested it was evidence that support of research pays off. But he was later forced to acknowledge that he had misread the health statistics—there were actually more than 31,000 cases of smallpox reported in 1970 to the World Health Organization.[33] The point seems to be this: Exaggeration in behalf of science (as when Handler exaggerated the impact of budget cuts and the achievements of health researchers) is permissible; exaggeration in a cause which leaders of the scientific community find threatening (as when Cole predicted disaster from our current fossil-fuel technology) is punishable by ostracism.

Incidents such as the barring of Cole seem relatively infrequent. In general, the Academy has elected competent scientists without regard to their public or political views. Its current membership includes Willard Libby, an outspoken "hawk," and Linus Pauling, a leading "dove." In recent years the Academy has elected Richard Lewontin, a radical geneticist who has led disruptions at national scientific meetings; Stephen Smale, a radical mathematician who publicly denounced the American and Russian governments while on a visit to Moscow and ended up in bitter conflict with Handler as a result of the incident; and Noam Chomsky, a radical linguist.

However, political and social considerations do affect Academy elections in the sense that some of the scientific community's own leaders appear to be elected more as a reward for loyal service in the top administrative posts of the scientific "establishment" than in recognition of outstanding original research. This is a point that is difficult to document. The list of Academy members who are probably better known for their administrative contributions than for their research might include, among others, the personable H. Guyford Stever, who was elected the year after he became director of the National Science Foundation, the government agency chiefly responsible for supporting scientific research at the universities; James A. Shannon, who guided the National Institutes of Health to major-league status as a research agency; S. Dillon Ripley II, suave secretary of the Smithsonian Institution; J. George Harrar, former president of the Rockefeller Foundation; and the late R. Keith Cannan, an influential division chief at the Academy itself.

But each Academician whose scientific credentials might be questioned inevitably has supporters who contend that his scientific work alone (without considering administrative accomplishments) justified election. Moreover, all candidates, even those nominated by the Academy leadership, must be ranked by the class committees, then voted on by the entire membership in the mail ballot—and it is impossible for an outsider to know just why the committeemen or the electorate rate a candidate as they do. Still, a candidate's public accomplishments clearly play some role, if for no other reason than that they make his name better known. Astin, who headed the National Bureau of Standards for many years, acknowledges: "If I hadn't been at the Bureau, I don't think I'd have been elected."

Some critics suggest that, if the Academy is going to elect people for "public service," it should broaden its definition of public service to include activist boat-rockers as well as servants of the scientific establishment. "Clearly in these days the area of public service ought to be considered as one component of a man's qualifications for election, but unfortunately this frequently works in the wrong way," complains Yale biologist Arthur W. Galston, who is not a member of the Academy. "Men like, for example, Barry Commoner, Paul Ehrlich and Lamont Cole, have not been elected, probably because there are some in the Academy who object to their political stance. On the other hand, there are others who get elected despite second-rate scientific credentials, simply because of their political or social position."[34]

If the Academy were nothing more than a highly exclusive club, we might agree with Academician W. H. Bradley that its election procedures are "strictly its own business."[35] But the Academy has another, more public function—the dispensing of advice to the government. And the quality of that advice depends, at least indirectly, on the quality of the membership. The members elect the officers and councillors who determine what tasks the Academy will accept; their attitudes, attention, and indifference establish the environment within which the officers operate; they sometimes suggest names of scientists who might serve on the advisory committees; they occasionally serve on such committees themselves (the Academy's most prominent committee, the Committee on Science and Public Policy, is composed entirely of Academy members); they serve as expert reviewers for many of the Academy's most important advisory reports; and they serve on newly established visiting committees which seek to determine how well the advisory apparatus is functioning.

Relatively few of the Academicians participate vigorously in the advisory work. President Handler lamented in April 1972 that "members of the NAS experience a serious lack of sense of responsibility and participation in the multitudinous and diverse affairs of the total organization."[36] But Handler has been trying to increase the involvement, so the quality of the membership will become even more important in the future. As of April 1974, some 403 members of the NAS—about 40 percent of the membership—were participating at least to the extent of agreeing to review an occasional advisory report.[37]

What the membership does—or fails to do—can have a significant effect on Academy operations. The Academy has been slow to endorse environmental concerns (see the chapters on pesticides, airborne lead, defoliation, and the supersonic transport). Can this be traced, in part, to the fact that there are so few ecologists and environmentalists in the Academy membership? The Academy has seldom challenged established power structures. Is this partly because the membership tends to be aged and somewhat "establishment"-oriented rather than young and iconoclastic? Some of the Academy's advisory units, such as the Building Research Advisory Board, have operated, at least until recently, with relatively little supervision from the Academy leadership. Is this because such units deal heavily with applied fields in which the Academy membership has little interest or aptitude? The Academy has seldom, if ever, questioned whether it is good public policy to concentrate research funds at the elite universities. Is this partly because the Academy itself is dominated by those universities? (Academician D. Stanley Tarbell contends "there is almost a conflict of interest" when "officers and members of the Academy, who represent the more favored universities," make recommendations to the government concerning research support. From the vantage point of Vanderbilt, a university in the hinterlands, Tarbell asserts: "It is not in the national interest to concentrate funds for research in a very small number of institutions."[38] Finally, the Academy often fails to recommend any solution to the problems it considers—instead, it simply recommends more research to help find solutions to the problems. ("I always knew what an Academy report would recommend," jokes J. Herbert Hollomon, former Assistant Secretary of Commerce for Science and Technology. "More research. More education of scientists. And, if the problem was terribly important, a national institution to cope with it.") Can these suggestions of research rather than solutions be traced, in part, to the fact that the Academy membership is composed of basic researchers rather than applied scientists?

The deficiencies of the Academy membership should not be exaggerated. Even those who find fault with the election procedures generally concede that the Academicians are, by and large, a distinguished lot. But they are not necessarily the best of American science, nor are they drawn equitably from the full range of scientific disciplines and institutions, nor do they include the young, vigorous, active scientists working at the frontiers of knowledge. Thus, for the sake of fairness, and in the interests of improving its advisory performance, the Academy should cast its electoral net much wider than in the past. At the very least, the Academy might open its nominating process to outsiders, a step which would seem relatively easy to accomplish administratively by allowing anyone to submit nominations, as the Pulitzer Prize competition does. But, beyond that, the Academy might also consider a mechanism whereby qualified outside scientists would be able to vote. This might be difficult to work out administratively, but it would do away with the electoral inbreeding which currently allows a handful of university departments to dominate the elections. If a fairer elections system can be devised, President Handler might eventually be justified in saying, as he did in a December 1971 letter to a Congressional committee, that the Academy membership "consists of the most distinguished 875 scientists in our nation in the fields of mathematics, and the natural, social, engineering and clinical sciences."[39] But as things stand now, that statement is an exaggeration.

CHAPTER THREE

How the Brain Bank Functions— And Malfunctions

"The National Academy of Sciences is the science brain bank of America. Its internal workings are as obscure to most people—indeed, to most of its members—as is the elaborate decision-making apparatus of the Vatican."

—John Lear, *Saturday Review*, July 5, 1969.

"One thing troubles me. I often read in the press that the Academy renders such and such advice. Sometimes the advice is quite contrary to my own judgment. This does not bother me at all, for I understand the way in which the Academy acts. But the public does not. And in this situation lies some danger."

—Vannevar Bush, head of the Office of Scientific Research and Development during World War II, in a letter to NAS President Philip Handler, December 8, 1969.

The lines of power in the Academy are hazy. The man who legally holds most of the power is the president. Under the Academy's constitution, he is authorized to accept any task requested by the federal government without bothering to get anybody else's approval (in practice, he almost always gets the approval of the Council or Governing Board), and he has the power to appoint committees. Thus he has the primary responsibility for determining what work the Academy will do and who will do it. "That's most of the power there is," comments one staffer.

The president also has immense powers to influence the wording of Academy reports. He sometimes sits in on meetings of committees that are handling particularly sensitive reports (under the Academy constitution he is ex officio a member of all committees that advise the government[1]). He is generally the court of last resort for settling internal disputes over the wording of a report. If there is a dispute, he sometimes rewrites a report, then seeks the committee's approval for his version. He determines whether or not the report will be transmitted. And he has the opportunity to attach a cover letter to all reports, which gives him a vehicle to express his personal opinions on an issue even if those opinions conflict with the judgment of the experts who have prepared the report.

These powers were all exercised in one of the Academy's most important recent reports—a 1973 evaluation of automobile emissions control technology by the Committee on Motor Vehicle Emissions. President Handler sat in on many committee meetings; he drafted a proposed section of the report; and, when the committee refused to accept parts of his draft, he published much of the rejected material in his cover letter, according to John E. Nolan, the executive director of the committee.

The cover letter included an estimate—not endorsed by the committee—that the cost of equipping all cars with the dual-catalyst emissions control system favored by the American automobile industry could eventually reach $23.5 billion a year, a figure of such magnitude that Handler was led to question whether this would be a wise use of funds for public health.[2] *The New York Times* even cited Handler's figure as justification for recommending that the automobile industry be granted a year's reprieve from the necessity to meet strict emissions standards mandated by the 1970 Clean Air Act Amendments.[3] However, two members of the Academy committee—Richard L. Garwin, a research physicist for International Business Machines Corporation, and James A. Fay, professor of mechanical engineering at the Massachusetts Institute of Technology—complained in a letter to the *Times* that Handler's estimate was "unrealistically high." The saw "little reason" to delay "a needed abatement program."[4] But their views received less attention than Handler's, largely because Handler's personal opinion could be expressed in an official cover letter and embued with the authority of the Academy presidency. So in this case the cost estimate emanating from the Academy was the opinion of one man, Handler, who was not an acknowledged expert in economics or automobile technology.

At the highest policy-making level, the only brakes on the president's power are the Council, a board of directors that sets policy and is the final authority on budgetary matters, and the Governing Board of the NRC, which has been delegated authority to establish committees, authorize and terminate contracts, and make certain appointments.

The Council consists of seventeen members—the Academy's five officers plus twelve other Academicians, all of whom are elected by the membership of the Academy. The group traditionally includes some of the top names in science and administration from universities and industry, with a handful of elite institutions dominating, as well as several former government officials. But the Council is seldom a major force in Academy affairs. It generally meets only six times a year, while its executive committee meets another six times annually. Moreover, most of the councillors are busy men whose main interests lie elsewhere; they are often poorly informed about details of the Academy's advisory work. "At the last Council meeting we talked about the doggonedest most trivial bunch of things," James A.

Shannon, former director of the National Institutes of Health, complained to the study team in a 1971 interview. "I finally said, 'We've discussed five or ten things and as far as I can tell it won't make a difference either way to me or to the Academy or to the scientific community.'" The Council seldom opposes an Academy president's policies; instead, it generally supports the president, much as the trustees of a university or the directors of a corporation tend to support the incumbent management.

The Governing Board of the NRC, which, under a 1974 reorganization, consists of seven representatives from the NAS, four from the NAE, and two from the IOM, may try to take a more active role in supervising the advisory mechanism. But the influence of the Academy leadership over the day-to-day activities of the organization is traditionally minimal. The very nature of the Academy—a far-ranging amalgam of part-time committees whose members are often prestigious and influential in their own right—makes centralized direction difficult. And, to compound the difficulties, the Academy is remarkably thin in top management personnel. For most of its existence, the Academy did not have a single full-time elected officer to head its large full-time staff. Then, in 1965, the presidency was made a full-time job; it remains the only top post at the Academy that is not treated as a part-time job to be filled by someone whose main interests lie elsewhere. The incumbent president, Handler, is the only elected officer who is paid. (In fiscal 1973, he received $67,500 [$50,000 in salary and $17,500 in deferred compensation]. Among other perquisites, he lives in an Academy-owned apartment, for which he is charged a modest rent, in the posh Watergate complex, along the Potomac River in downtown Washington, D.C.; and he is provided with a chauffeur and limousine for official duties.) Although Handler is considered an articulate spokesman for the scientific community, his close assistants question his management skills—he had never previously run anything larger than the department of biochemistry at Duke University Medical Center. The remaining four elected officers—vice-president, home secretary, foreign secretary, and treasurer—serve on an unpaid, part-time basis. The Academy has no full-time legal counsel—and it only acquired a full-time comptroller as recently as 1969.

Considerable influence over the organization's activities falls by default to the permanent staff, headed by Executive Officer John S. Coleman, which is on the scene at the Academy's Washington headquarters every day. The staff plays the major role in formulating proposals for projects and in selecting the experts to carry them out. It also provides administrative support for the experts who conduct the studies, a circumstance which gives a few aggressive staff members the opportunity to influence the wording of Academy reports. In mid-1974, the staff totaled 1107 persons, of whom 109 held degrees at the doctorate level.

The professional caliber of the staff—its scientific and technical ability—has never been one of the Academy's strong points. A few staffers are considered highly competent by their professional peers but most are considered mediocre. A high Pentagon science administrator told us: "The staff is not very impressive in my view. Most of them rank in the lower 50 percentile of technical people. It's the tired old bureaucrat kind of thing." A member of the NAS Council told us: "Very few of the staff are first-class people." And J. Herbert Hollomon, former Assistant Secretary of Commerce for Science and Technology and a founder of the National Academy of Engineering as well, told us the staff is "not much good." Some Academicians believe the staff is as good as it need be, given the routine nature of the tasks it performs. One experienced veteran of the Washington science advisory scene even believes the staff compares favorably with that of Congress. Some staffers are competent enough professionally to make a substantive contribution to the work of their committees, but most do little more than handle bookkeeping, correspondence, travel arrangements and other administrative chores. They are "supporters" rather than "contributors." As of 1974, the staff remained one of the weak points of the Academy.

The actual work of the Academy—the writing of advisory reports and framing of recommendations—is done largely by outsiders who are subject to only minimal guidance from the Academy leadership. The key working unit is the committee—either an ad hoc committee appointed to deal with a particular issue (as the Committee on Motor Vehicle Emissions, which was appointed to deal with the question of whether the automobile companies could meet pollution standards mandated for 1975) or a standing committee which renders advice on a subject year after year (as the Committee on Food Protection, which continuously monitors problems of toxic substances in the food supply). The members of these committees are drawn from the scientific community at large—from the universities, industrial laboratories, government agencies, or other institutions. They meet periodically, perhaps once a month or even less frequently, to discuss the problems assigned to them. And they generally draft their own reports, though in some cases a staff member will write the first draft and submit it to the committee for revision.

The selection of the experts who will serve on a committee is usually the most crucial factor in determining how good a job the Academy will do on any given project. The key role in appointments is played by one of the major operating units of the Academy—an assembly or commission, a division, a standing board or committee, or other such units. The staff, or perhaps the outside scientist who chairs the unit, solicits suggestions from a variety of sources within and without the Academy. The process depends heavily on personal, word-of-mouth recommendations, a circumstance which insures that all candidates are fairly well-known quantities but which tends to exclude people from outside the Academy's immediate circle. The appointments process has been aptly described as a "buddy system"—even by one of the Academy's own reports.[5]

After the relevant operating unit has developed a list of proposed committee members, the names are screened by the staff of the president's office in an effort to determine whether the committee seems competent to perform the job assigned and is reasonably balanced or impartial in outlook.

As the jobs thrown to the Academy become increasingly complex, the amount of time demanded of committee members seems to be increasing. This service is generally donated by committee members without charge. Individuals who work on a project on a sustained basis, for many weeks, as in a 1970 summer study of Kennedy Airport's expansion plans, are paid for their work, but the vast majority of committee members work on a part-time basis and receive no fees, merely travel expenses. The dollar value of this contribution is considerable. If the same individuals were serving on an advisory committee to the National Science Foundation, they would receive expenses plus a fee of $75 to $100 per day. If they were consulting for industry, they would almost certainly be paid even more, perhaps $200 to $500 a day. Academy president Handler estimates that the total value of the consulting services contributed without charge is roughly $6 to $7 million a year.

No one has done a systematic survey of the motivations involved, but it appears likely that a variety of factors influence individuals to serve on Academy committees. These include a desire to serve the nation, a sense of obligation, the opportunity to rub shoulders with distinguished colleagues, the opportunity to pick up information that might be useful in one's professional work or business, or the hope that meritorious service on committees might ultimately lead to election to the honorary Academy itself. That hope is apparently well-founded: Of 168 new members elected to the honorary NAS in 1972 and 1973, fully seventy-three had served on Academy advisory committees during the previous decade.[6] There is also a tradition that academic scientists devote their free time to a variety of "professional duties." And, for some, the lure of "teaching" the government what it should do proves irresistible. "The 'professor' whose advice is requested—whether he comes from academia, industry, or a government laboratory—finds among his students Cabinet officers and Congressmen, generals and White House aides," exults one Academy publication.[7] But, whatever the motivation, the important fact is that most of those who are asked to serve gladly do so, and many contribute substantial labor to advising on public issues.

Unfortunately, the reports developed by the committees are often flawed. There is a saying passed around among Academy staff members to the effect that 5 percent of the Academy's reports are superior, 5 percent are inferior, and 90 percent are of a passable but pedestrian quality that reflects little credit on the Academy and does little good for the nation. The percentages vary with the teller, but the point is generally the same; one of the biggest problems facing the Academy is that so few of its reports are of the very highest quality. There appear to be several reasons why this is so.

One is that the Academy does not always appoint the very best experts to handle its advisory tasks. All too often it puts together committees which are not the most qualified to deal with the problem presented. This happens largely because the Academy's "buddy system" tends to draw talent from a relatively narrow slice of the scientific community. Some committees are dominated by small, inbred cliques that are out of touch with differing viewpoints. And others lack the expertise needed to consider an issue fully.

The quality of Academy committees seems to vary haphazardly. Lewis Branscomb, a member of the Academy Council who once headed the National Bureau of Standards and is now vice-president and chief scientist at IBM, told us he knows of some committees which are made up of "energetic younger guys," while others "go on and on with well-meaning nice guys who have dropped from the main stream of what's happening." Academy Vice-President Kistiakowsky lays much of the blame for poor committees on the "undesirable" influence of the Academy staff members who play such a big role in the selection process. "The committees are sometimes not made up of the best possible people [because] the bureaucracy doesn't know them," he says.

The bureaucracy also tends to avoid "controversial" appointments with the result that committees are often dominated by bland "Establishment" types, while shunning the activist boat-rockers. When the Academy appointed a committee to assess the biological effects of ionizing radiation, for example, it ignored the two scientists who had sparked the public outcry against radiation hazards—John W. Gofman and Arthur R. Tamplin, of the Livermore branch of the Lawrence Radiation Laboratory—but it appointed the scientist who had played a major role in rebutting Gofman and Tamplin on behalf of the Atomic Energy Commission, Victor P. Bond, associate director of Brookhaven National Laboratory.

The Academy never officially explains the factors which determine the selection of a committee, but private conversations with Academy staffers indicate that activists are often excluded on the grounds that they are "biased" or lack the necessary expertise or are too "abrasive" to work effectively on a committee. Whatever the reasons, the end result is that Academy committees tend to be dominated by "safe" scientists who will not push too hard to change the status quo.

The committees also tend to be middle-aged. In fiscal year 1970, the median age for members of Academy advisory committees was fifty, which was about ten years older than the median age for all doctorate-holding scientists in the nation. Less than 3 percent of the Academy advisers were in the thirty-five-and-under age bracket.[8] Academicians sometimes justify the age of their committees by asserting that advisory tasks require experience and judgment. That is partly true, but the Academy makes too little use of young scientists whose vigor is not yet diminished by the passage of time and whose willingness to challenge established ideas is apt to be high. Women and minorities also seem underrepresented, constituting 3.2 percent and 2.3 percent of the membership of Academy committees in April 1974, but their numbers have been rising steadily in recent years.[9]

Even when a committee consists of individuals who are well qualified in their specialties, it sometimes lacks the breadth of vision needed to grapple with the problem assigned. This is particularly true when the issue involves social or economic problems which lie outside the Academy's traditional realm of expertise in the natural sciences. In 1968, for example, the Academy's Committee on Problems of Drug Dependence co-authored a statement, along with two committees of the American Medical Association, on the dangers of marihuana. The statement branded marihuana a "dangerous drug" and a "public health concern." And it called for more flexible laws that would distinguish between casual offenders of the marihuana laws and chronic users or dealers. But it opposed legalization of marihuana on the ground that such action would "create a serious abuse problem in the United States."[10] This stand was perceptively attacked by John Kaplan, professor of law at Stanford University and an advocate of legalization, on the ground that the medical experts on the Academy committee had failed to consider the many complex social issues involved. Kaplan acknowledged that marihuana is in some sense dangerous, but he questioned "whether the state should incur the social costs of making people do what is felt—even correctly—to be good for them." Indeed, he suggested that "marijuana might be dangerous in the medical sense and yet, unlike heroin or LSD, not be so dangerous as to justify the costs of criminalizing it."[11] That possibility was not even discussed by the Academy statement.

In an effort to eliminate such lacks in the future, the Academy began, in the early 1970s, to elect more behavioral scientists to its membership and to appoint experts from such diverse fields as law, business, and political science to some of its advisory committees. But it continued to show remarkable blind spots. A study group which issued a 1973 report on how American corporations could help strengthen the technical capabilities of developing countries was dominated by American businessmen—it included not a single representative from a developing country. The imbalance was so blatant that even the committee acknowledged that "it could be accused of a one-sided view of strongly contended issues."[12]

The conditions under which Academy committees work is another major impediment to performing a good job. Even the most highly qualified experts find it difficult, if not impossible, to solve perplexing national problems in a few harried days of committee meetings. This is not generally appreciated by the general public, which tends to regard Academy pronouncements as the ultimate wisdom, but it is readily acknowledged by many Academy leaders.

The problem of part-time-itis is compounded when committee members fail to do their homework before coming to the meetings. One FDA official who has served as liaison to several Academy committees told us he is amazed at how some of the distinguished experts will arrive at a meeting unprepared to discuss the subject at hand. And the problem is aggravated when committee members miss some of the meetings because of obligations elsewhere. The Academy keeps its absentee records hidden, but the one committee for which the study team obtained almost complete minutes— the 1969 Committee on Persistent Pesticides—showed a substantial absentee rate. One member of that committee missed seven of eleven meetings for which we have records, two others missed four meetings, and seven others missed from one to three meetings. Only five had perfect attendance.

An unfortunate by-product of the part-time committee approach is that the Academy often responds too slowly to meet the needs of the government. In 1969, for example, the Academy received a grant from the Justice Department to review genetic knowledge on the XYY chromosome pattern in man. The legal agency was interested in the matter because recent studies had suggested a link between the XYY chromosome combination and delinquent or antisocial behavior. But the Academy was so slow in getting started that "the grant was discontinued at the request of the Department of Justice because of other active research projects undertaken to answer the problems in that area. No meeting of the committee was held, and it was discharged."[13]

The Academy record for slowness in completing a task appears to be held by a committee that compiled a comprehensive report on the 1964 Alaska earthquake and its consequences. The project was launched some ten weeks after the quake took place, and it was originally expected to take no more than three years. But the project was slowed by wrangling with the government over how large a budget was needed, uncertainty about what form the report should take, difficulty in coaxing manuscripts out of panelists who were not, after all, being paid for their work, lack of an adequate editorial mechanism, frequent financial crises, and other difficulties. The final report was not completed for almost a decade, long after the quake had faded from the memories of the public and government bodies. Even the committee chairman acknowledged that, "in any rational view, it seems incredible that a disaster report in this day of speed and efficiency should require for completion nearly a decade following the event."[14]

There appear to be inherent limits to what any committee is apt to accomplish. To begin with, committees are seldom a source of original thinking. As one of the Academy's own reports acknowledges: "New ideas for fruitful research or the solution of technical problems usually have their genesis in the minds of individuals, not in the deliberations of groups."[15] Moreover, committees tend to avoid controversy and disagreement by seeking a consensus, with the result that most Academy reports come out rather bland and unexciting.

A further problem, somewhat peculiar to Academy committees, is that they often view their job more as an educational experience for the members than as an exercise in problem-solving. One Academy committee actually scheduled time at each of its meetings for "discussion of some scientifically interesting and relevant topic,"[16] while a survey of Academy pesticide committees revealed that the members were more interested in the "learning experience" and the prestige of being appointed to an Academy committee than they were in affecting public policy.[17] That sort of approach may be valuable for the committee members, but it often makes Academy reports less operationally useful than they might otherwise be. Government administrators repeatedly complained to the study team that Academy committees "have their heads in the clouds" and often fail to produce reports focused on the needs of agency officials who must take action on a problem.

Documentation for this mismatch can be found in a review carried out by the Department of Agriculture. As a result of our inquiries, Ned D. Bayley, then Agriculture's Director of Science and Education, became curious as to just what use the Academy's services had been. He ordered a formal review of all contracts between the Agricultural Research Service and the Academy from fiscal years 1968 through 1972. The review, which was conducted by Harry W. Hays, who had once worked for the Academy, came up with surprisingly negative findings.[18]

Hays found that when the Academy proposed a project and Agriculture agreed to fund it, the scope of the undertaking was generally "so broad that no committee could possibly make an in-depth study of all the facets outlined in the proposal in the allotted time." As a result, the Academy failed to deliver many of its reports.

The results were also unsatisfactory in cases where Agriculture originated a contract and asked the Academy to study a specific problem. "In reviewing the contracts," Hays said, "it would appear that the Academy took the initiative to set up a plan of study that was in accordance with its procedures and approach to scientific problems rather than to meet the needs of ARS."

Not all of the blame for these misfires should necessarily be placed on the Academy. Hays acknowledged that "The fault could also have been due to our failure to make our requirements absolutely clear in the charge to the committee." Moreover, once a contract was let, Agriculture does not seem to have paid much attention to whether the Academy was doing anything useful. Hays found that "in some cases" Agriculture's contract representative "was not fully knowledgeable about the contract, the status of the study, progress reports, or what ARS did with any recommendation the committee might have made." Some Academicians complain that the Department of Agriculture is not exemplary enough to pass judgment on the prestigious Academy. But Agriculture's survey of the Academy's performance appears to be the most systematic carried out by any agency in recent years, and it seems evenhanded in its approach, laying part of the blame for failures on Agriculture itself.

Until recently there was very little quality control over the output of Academy committees. But under Handler's presidency the Academy has established an elaborate review procedure designed to insure that virtually every report the Academy produces is read and criticized by a group of scientists who have had no hand in preparing the report. The reviewing process must be considered a major accomplishment of Handler's administration. Previously, if an Academy committee fell behind the times or became biased, there was little the Academy could do to monitor the quality of its work. The committee could continue to issue reports in the Academy's name, year after year, no matter how wrongheaded or biased those reports might be. But with the formation of the review system the near autonomy of the committees was eroded a bit. Each committee was henceforth on notice that its report might be scrutinized by distinguished scientists who would pass judgment on whether it was clear, concise, convincing, complete, and fair, or marred by conflict of interest. The reviewers are not generally supposed to "second-guess" the authoring committee on matters of technical substance, for the authoring committee is supposedly made up of the best technical experts who can be found. Rather, the reviewers are supposed to judge whether the report is internally consistent and balanced, and whether its conclusions flow logically from the data presented. The system is by no means foolproof. Some reviewers do a perfunctory job; others are too timid to challenge the authoring committee; while still others assume an arrogant posture and try to impose their own biases on the authoring committee. Moreover, the reviewers sometimes have little effect on a report even where their criticisms seem to have been correct.

The review process is purely advisory; the final decision on what a report will say remains with the originating committee. This is explicitly enunciated in a notice attached to the front of most Academy reports which states that "responsibility for the detailed aspects of this report rests with that committee." If the reviewers and the originating committee are locked in irreconcilable disagreement, the issue is passed to the Academy president for resolution. But, if an originating committee refuses to back down, the president cannot compel it to change its report. The most he can do is refuse to transmit the report, or transmit the report with a cover letter disavowing it. Either course of action involves public embarrassment and friction within the Academy complex. Moreover, if the Academy fails to deliver its report, there may be a financial penalty as well, for the Academy may have to return the money it has received under government contract. The usual result when there is sharp conflict between the originating committee and the reviewers seems to be that the originating committee agrees to changes in wording to placate the reviewers, but the basic thrust of the report remains unchanged.

It is hard to see how the Academy could do otherwise than leave the final responsibility for a report up to the expert committee it has assembled, for it would seem dangerous to grant a few high officials of the Academy power to impose their own views on the contents of the hundreds of reports produced annually, most of which lie outside their areas of expertise. But readers of Academy reports—particularly public officials who plan to act on their recommendations—should be aware that they are not really Academy pronouncements. The reports are the work of small committees of experts whose performance can be flawed by the host of factors enumerated in this study.

CHAPTER FOUR

The
Special Interests

"The Academy of Sciences is unique among scientific institutions in the United States. Here is the pinnacle of achievement and competence, the ultimate in objectivity and status. The Academy brings together the best scientific talent of our Nation to serve as one body—without conflict of interest or bias—in providing advice to our government on scientific matters."

—T. C. Byerly, former chairman of the division of biology and agriculture, at the annual banquet of the division, March 1969.

"One thing that is missing is a credible group which can lay out in terms understandable to the public, Congress, and the executive branch too, what the scientific and technological facts are and to do it in an unbiased and credible way. The National Academy's function comes close to this, but it may not be completely adequate simply because it operates under federal charter. I believe we need groups which can speak in an unbiased, straightforward way without the kind of adversary relationships which you inevitably run into with these complex questions."

—Edward E. David, Jr., science advisor to former President Nixon and a member of both the National Academy of Sciences and the National Academy of Engineering, in an interview with *Mosaic*, official publication of National Science Foundation, Summer 1971.

The Academy claims that the most distinctive feature of its committees is that they are independent of any pressures from special interests. The Academy supposedly provides a neutral enclave where experts can say what they really think without fear of adverse reactions from government, industry, or any other sources; and individual members of Academy committees are supposed to offer their best opinions as professionals rather than as spokesmen for any particular interest group. The end result, if one were to believe the ringing words of former Academy president Frank B. Jewett, is that the Academy is "A Supreme Court of final advice" whose findings are "wholly in the public interest uninfluenced by any elements of personal, economic or political force."

But the Academy's record in recent years suggests that its protestations of Supreme Court impartiality should not be taken at face value. In actual practice, many of the Academy's reports have been influenced by powerful interests that have a stake in the questions under investigation. An unusually frank admission of this came from Harvey Brooks, one of the most influential Academicians, in an interview with the *National Journal*. "It's true," Brooks conceded, "that some of our bodies—the Highway Research Board, the Food and Nutrition Board, the Building Research Advisory Board and the Space Science Board, for instance—may be constituted too completely with those who have an economic or institutional interest in the outcome of their work."[1]

We found no cases of direct, personal conflict of interests at the Academy—no cases, for example, where a committee member profited financially as a direct result of the advice he rendered. This does not mean that such cases do not exist. The Academy is such a sprawling institution that it is difficult to make any categorical statement about it. But Academy Vice-President Kistiakowsky asserts that the typical conflict of interests involving financial shenanigans "just doesn't exist here, not at all." However, there are cases where committee members render advice that has an indirect effect on their professional and economic interests. Thus, committees of scientists routinely recommend greater government support of their particular discipline, a recommendation that would increase the pool of funds from which their own research is supported. And scientists on Academy committees often render advice that could affect the economic fortunes of their employers, as when an industry scientist serves on a committee considering pollution standards.

Moreover, even without any conscious consideration of the economic stakes involved, Academy committees often suffer from a generalized form of bias in which committee members are influenced by the viewpoints of the institutions and professional worlds within which they routinely operate. It would be unfair to suggest that a scientist's background automatically condemns him to reach a particular conclusion on an issue. Such an assumption would suggest that all human beings are slavishly programmed by their past experience, with no room for independent judgment and willingness to accept new facts and new viewpoints. Nevertheless, one can often predict how a scientist will lean on a particular issue simply by knowing his background. An aerospace engineer, for example, is apt to favor the space shuttle whereas a biologist might be skeptical; a food technologist is apt to favor the use of chemical additives whereas a geneticist might view them as dangerous; an agriculturalist will probably favor the use of pesticides to protect crops whereas an ecologist might disapprove; and a military technologist is apt to favor a new weapons system whereas a disarmament specialist might be opposed. Each of these experts is apt to push for policies that are consistent with his professional outlook and that will most benefit the institutions within which he operates. As Brooks has described the tendency:

> It is notorious that chemists from the chemical industry and engineers from the auto industry were slow to accept the evidence of environmental pollution resulting from their products. This was not out of greed, or the profit motive, or even "just obeying orders from above," as some now seem to be

asserting. The skepticism was usually quite sincere and consciously disinterested even when subconsciously biased. It is difficult for anyone to accept criticism of his brain children, and this may be particularly so in the case of engineers or technologists who may have committed a large slice of their careers to a single goal.[2]

This form of bias is difficult to deal with, for it is often unrecognized by the individual, by his colleagues, and by the general public. But it is a very real problem that often undermines the objectivity of Academy reports. And beyond the problem of individual bias, there are a host of problems—structural, financial, and legal—which impede the Academy's efforts to offer "independent, objective, untrammeled" advice. In the remainder of this chapter, we shall examine how the Academy sometimes falls prey to pressures from three special interests—the government, industry, and the scientific community.

The Academy claims to be both a private organization and an officially designated adviser, a combination which supposedly enables it to operate close to government but with less concern for the prevailing political winds than is possible for the government's own in-house advisory committees.

But how independent are the Academy's own committees? Are they really free to initiate and criticize? A close inspection reveals that, while they are not an integral part of the federal government, they are nevertheless subject to government influence in many ways.

The most important government control is the power of the purse. The Academy has relatively little money of its own; the great bulk of its work is financed by government contracts. Thus it is difficult, if not impossible, for the Academy to undertake a major study, particularly a costly one, unless a government sponsor can be found. The government, in effect, can veto any study that looks potentially troublesome to political leaders. The pernicious consequences of this situation became apparent in 1971 when the Academy volunteered its services to conduct a study of the ecological consequences of building the proposed Trans-Alaska Pipeline, a multibillion-dollar project to transport oil from the Arctic down to a port in southern Alaska. Interior Department officials thanked the Academy for its interest but indicated that they already had all the scientific advice they needed.[3] The project was apparently just too big, and too politically sensitive, to risk letting the Academy make recommendations that might conflict with Administration plans to have the pipeline built. And the great cost and logistical problems involved in studying an area of such size made it virtually impossible for the Academy to mount a study using only private resources. Congressman Les Aspin (D-Wis.) later said that the government's refusal to help fund the Academy study "raises very grave suspicions about whether Interior believed its conclusions in favor of an Alaska pipeline could stand up to close scrutiny."[4]

The government's financial powers also affect the Academy's handling of projects which *are* funded and carried out. It would be overstating the case to say that the government "buys" a favorable report from the Academy, or that "he who pays the piper calls the tune." But the Academy's dependence on the federal government for financial support has a dampening influence on the Academy's willingness to be sharply critical, for if an Academy committee gets too nasty about an agency's programs, that agency is unlikely to continue supporting the committee's work. The most dramatic example we found of this problem involved a committee which advised the Atomic Energy Commission on radioactive waste disposal. When that committee got too belligerently critical of the AEC, the agency cut off its funding, and the Academy was forced to set up a new committee more in tune with the AEC's own thinking.

That was an unusually blatant instance of agency retaliation and Academy capitulation, but the problem is a general one that affects virtually all of the Academy's standing committees and staff members. These committees and staffers are largely supported by federal contracts, and when one contract expires there ensues a scramble to obtain another contract to replace it, lest the committee have to disband and whatever influence it has come to an end. In that kind of atmosphere, it is not surprising the committees hesitate before saying anything that will antagonize a sponsoring agency.

An unhealthy symbiotic relationship often develops between the advisory committees and their federal patrons. The federal agencies keep some committees alive by assigning them trivial work, and the committees keep the agencies happy by handling the few genuine problems thrown their way in a docile manner. Academy officials seldom acknowledge that their committees try to remain in the favor of funding agencies, but they recognize that there are serious problems in the relationship. Harvey Brooks told the study team that staff members, whose salaries are financed from government contracts, sometimes "see their future career as dependent on pleasing the agency" and sometimes "try to persuade the committee to slant its reports to please the agency." But Brooks said the committee members, who earn their salaries elsewhere and are not paid for their committee work, are not generally susceptible to the same pressures that afflict the staff.

Handler, in a private letter to Academy members in 1972, also acknowledged the "internal hazards" of the Academy's financial dependence. He noted that, when a committee created to perform an advisory task for the government completes its job, it then has

> gained a certain coherence, adopted some life style of its own, and even developed a constituency. But then, no formal business, no government request is in hand. . . . But the committee may wish to remain alive, considering that it has a unique and useful role to fulfill, and the staff finds itself caught up in the endeavor. If our funds permitted, if the committee had demonstrated its competence, and continuation seemed desirable, we might find it appropriate to maintain the committee at some relatively low-cost, standby level so that it would be available in the event of a legitimate and significant request for advisory services. But we cannot adopt that course, for lack of our own internal funds. Due to this circumstance, the committee, its staff, or a federal agency may propose a relatively trivial project which can readily be financed, in which case the committee can remain alive.

Handler said he did not know "how frequently that alleged scenario has occurred," but he acknowledged that "it is said to be real" and "can pose troublesome questions."[5]

Although Handler did not specify just what all the troublesome questions were, a report on technology assessment issued in 1969 by the Committee on Science and Public Policy (COSPUP), the Academy's senior committee, suggested that NRC committees often let their reports be colored by the viewpoint of the funding agency. As COSPUP expressed it: "Any human institution has tendencies that, unless counteracted, will over time cause it increasingly to be run for the benefit of people inside the organization and for those special outsiders with whom they have found it easiest to identify themselves." COSPUP warned that if "an ostensibly neutral" assessment body "is financed by, works closely with, or reports to an agency with a promotional or restrictive mission of its own (as is ordinarily the case, for example, with the committees of the National Research Council), the assessment may be colored by the agency mission."[6] Consequently, COSPUP expressed doubts about the wisdom of allowing the Federal Aviation Agency to support an Academy study of the sonic boom, the Defense Department to support an Academy study of the impact of defoliation on Vietnam, and the Agriculture Department to support a study of the role of fertilizers in contributing to lake pollution; it recommended that, in the future, such studies be supported by an organization responsible for technology assessment rather than by an agency which is interested in getting a particular verdict from the Academy.[7]

Financial considerations of another sort—the Academy's desire to coax greater support for science from the government—also inhibit the Academy's willingness to criticize or oppose the government. In an interview with the study team, Academy President Handler indicated that there is a quid pro quo attached to Academy advisory services. The Academy, through its advisory committees, helps the government in the belief that a grateful government will thus recognize the utility of science and will be generous in its support of science. Handler noted that the scientific community is continually faced with the "burden" of arguing the case for federal funding of science. "And in that sense, we are welcome to make those arguments at the appropriate times and places by virtue of the fact that we are useful and earn our keep, as it were," Handler said. "If indeed we have performed useful services for the various agencies in government, then the size of the welcome mat when we want to make some arguments to a given agency with respect to science itself . . . that welcome mat is larger."

Such considerations almost certainly limit the Academy's freedom to operate independently of the government. In early 1972, for example, Handler addressed his membership on the question of whether the Academy should refuse to advise the military on projects which might contribute to the arms race. He warned that while such action "might be seen as a bold gesture taken by the Academy in the hope of dramatically signaling the need for international action to halt the arms race . . . it could instead prove to have been an empty, feckless gesture; the Academy, as we have known it, might be seriously damaged, perhaps destroyed, and with it, national and governmental respect and support for science injured as well."[8]

Legal constraints are another factor which renders the Academy subservient to the government. The Academy's charter, granted by Congress in 1863, states that "the Academy shall, whenever called upon by any department of the Government, investigate, examine, experiment and report upon any subject of science or art."[9] Although the Academy's responsibilities have never been defined in a court test, that clause seemingly requires the Academy to perform any task requested by the government. In practice, this requirement is not absolute. The Academy has, on occasion, turned down government requests, generally on the grounds that it was not competent to do the work or that someone else was better qualified to do it. Moreover, the Academy, in the course of negotiating a contract with the government, will often modify the government's original request somewhat. But for the most part the Academy, although nominally a private organization, feels called upon to do the government's bidding. A succession of Academy presidents has espoused this viewpoint. Jewett called the Academy "a creature of the State."[10] Seitz interpreted the charter to mean that "we should not arbitrarily close the doors of the Academy to portions of our government that come to us seeking advice in areas where we are clearly able to give it."[11] And Handler has said that, while individual Academicians are free to refuse to serve on advisory committees for whatever reason, the Academy as an institution has no such freedom. "I cannot see the Academy, as such, declining unless there is a genuine question about the legality of the government or the request itself is unreasonable or repugnant." (He suggested that the Academy would probably not help a Ku Klux Klan–oriented government official devise better ways to place blacks in concentration camps, for example.[12])

The combined result of the Academy's dependence on federal funding and its legal responsibility to assist the government upon request is that the Academy has become a technical handmaiden to the government. Federal officials can set policy without bothering to consult the Academy; they can then ask the Academy to help advise them on the best way to carry out that policy; and the Academy generally pitches in to help without worrying about whether the policy is misguided. Thus the Academy was never asked for its opinion as to whether the country should send a man to the moon, or build a supersonic transport, or set up an antiballistic missile system. Instead, it was simply asked to consider such questions as what scientific experiments should be conducted in space, whether the SST's sonic boom could be alleviated, and how best to harden ABM missile sites. No one would argue that the Academy itself should set policy on large national issues, many of which involve socioeconomic considerations outside its traditional competence. But the Academy's inability to say no to the government, coupled with its lack of funds to finance projects of its own, means that the Academy almost invariably plays a supportive, rather than challenging, role toward the government. Some critics would put it more strongly.

"The NAS is an instrumentality of the administration, giving its unquestioned (and sometimes blind) service to the policies of whichever political forces are in power,"[13] complains Richard C. Lewontin. That is probably overstating the case. But there is no doubt that the Academy's independence is circumscribed by its history, traditions, financial dependence, and charter.

The detailed conduct of an Academy study is supposed to be insulated from government control. But in actual practice, even this independence is eroded in subtle ways. The contracting government agency decides whether a project will be undertaken (most Academy studies are in response to government requests, and those that aren't must be acceptable to a funding agency or they won't get funded); it plays a major role in defining just what questions the Academy is to answer; it often proposes candidates to be committee members; it sends a liaison man to committee meetings (except for executive sessions); it often supplies the bulk of the data to be considered by the committee; and it often gets a chance to read and comment on a report before it is released. Agencies have sometimes even been able to suppress a report, though Academy officials claim that is a thing of the past.

All these factors give the government significant influence over Academy operations. To begin with, at the stage where a contract is being signed, the government often shackles the Academy by setting narrow limits to an advisory task. One instance of this came to light in 1972 when the Academy produced a report on the technical feasibility of Project Sanguine, a controversial communications system proposed by the Navy for sending messages to submerged submarines. The heart of the Sanguine system, which was then undergoing feasibility studies at a site in Wisconsin, was to be a huge underground antenna that would provide a supposedly indestructible means of communicating with missile-carrying submarines in the event of a nuclear attack on the United States. The Sanguine system aroused intense opposition in Wisconsin on military, environmental, and technical grounds. The technical objections, which were raised by a handful of scientists at the University of Wisconsin and elsewhere, involved complicated calculations which purported to show that Sanguine would not actually perform as well as the Navy was claiming.

As the opposition mounted, Senator Gaylord Nelson (D-Wis.) asked the Navy to get an objective evaluation, and the Navy in turn contracted for an Academy review of the issue. Although Nelson had asked for a "review of the Sanguine concept" as well as of the specific criticisms that had been made of it,[14] the Defense Department simply asked the Academy "to review two reports on Sanguine which challenge the validity of this project."[15] The Academy appointed a panel which reviewed the papers critical of Sanguine, examined much of the Navy's own data, interviewed the principal figures, and finally issued a report in May 1972 which essentially concluded that the critics had not made a persuasive case that Sanguine would not work, while the Navy's data suggested it probably would work. "The panel has found no reason to disagree with the technical methods used by the Navy for prediction,"[16] the Academy report said.

This conclusion was given added weight when it was endorsed by one of the Academy's most eminent members, Nobel Laureate E. M. Purcell, who was asked to review the report before it was released because of his expertise in the technical areas at issue. Purcell noted that the Academy panel had concluded that the critics were "demonstrably wrong" on the main technical points they had raised, and he added: "I do not see how the panel could have reached any other conclusion."[17] Thus the Academy, in its wisdom, had concluded that a handful of scientists, working nights and weekends and without access to all of the Navy's technical data, had failed to make a prima facie case that Sanguine wouldn't work. But left unanswered were such major questions as whether Sanguine was a desirable project; whether it was the best way to solve the problem of communicating with the nuclear submarine fleet; and whether it would have an adverse impact on the environment or on other means of electromagnetic communication. The Navy had been careful not to ask those crucial questions, and the Academy was thus largely limited to reviewing the criticisms of the project rather than the desirability of the project itself.

Another constraint on the Academy's independence is the interchange of personnel between the government and the Academy's advisory apparatus. Academy presidents traditionally sat on the U.S. President's Science Advisory Committee until that body was disbanded in early 1973. In that capacity, it was difficult for them to voice public opposition to administration policy on issues that were considered by PSAC without opening themselves to the charge that they were misusing confidential information they had received as presidential advisers. President Handler also sits on the National Science Board, the policy-making body for the National Science Foundation, while his predecessor, Frederick Seitz, chaired the Defense Science Board, the Pentagon's highest science advisory group.

The flow of personnel also runs from government to the Academy. A number of former government officials have been active in the Academy at all levels, from the Council to the staff to the part-time committees. In some cases, the Academy allows current or past government officials to hold positions in which some influence on Academy recommendations affecting the agency that employed them is possible. Thus, Irl C. Schoonover, who retired as associate director of the National Bureau of Standards in 1968 after forty years of service, subsequently directed a group of Academy panels which evaluated the research and technical programs of the Bureau; he even had his office at the Bureau's headquarters in suburban Washington, far from the Academy's own offices. T. C. Byerly, a scientific administrator with the U.S. Department of Agriculture, chaired the Academy's biology and agriculture division in the mid-1960s, a position which gave him powers over the division's many reports that impinge on USDA policy. The staff of the Maritime Transportation Research Board was headed by a retired Navy rear admiral, E. G. Fullenwider, from 1954 to 1968, and by a retired Coast Guard rear admiral, John B. Oren, from 1968 until recently, thus bringing the views of the uniformed services into the deliberations of committees which make recommendations affecting both civilian and military maritime policies. And the Academy's medical sciences division was headed from 1967 to 1973 by Charles L. Dunham, former director of the division of biology and medicine of the Atomic Energy Commission, who thus had jurisdiction over committees advising the government on safe radiation levels. Meanwhile, the National Academy of Engineering in 1973 elected as its president Robert C. Seamans, Jr., who had just completed four years of service as secretary of the Air Force and who had previously been a high-level administrator at the National Aeronautics and Space Administration.

Not every former government employee is necessarily a strong partisan of his former agency's interests, but the presence of agency alumni at the Academy increases the likelihood that Academy committees will fail to exert critical independence and will simply rubber-stamp agency proposals. That this is a real danger became apparent in a virtually unnoticed incident in 1969 when the Academy's medical sciences division, which is heavily populated with retired military personnel, joined a Defense Department effort to save a discredited medical education program. The program—known as Medical Education for National Defense (MEND)—provided some $1 million in federal funds to ninety-two medical schools in fiscal year 1968 on the understanding that those schools would give their medical students specialized training in military and disaster medicine. But in March 1968 the General Accounting Office issued a critical report which contended that the program had been mismanaged and wasteful,[18] and in July 1968 the House Appropriations Committee deleted the program from the fiscal 1969 military budget,[19] an action which was subsequently upheld by Congress.

Less than a month after the program had been killed, however, the Pentagon began laying the groundwork for an attempt to revive it. On August 8, 1968, the late Louis M. Rousselot, then deputy assistant secretary of defense for health and medical affairs, formally asked the Academy's medical sciences division to conduct a "critical evaluation" of MEND.[20] Rousselot told the study team in a 1971 interview that the Academy study was requested "to hopefully back up our strong belief of the importance of the MEND program." Rousselot could afford to be optimistic. He knew the evaluation would be in sympathetic hands. The Academy's medical sciences staff was headed by Henry T. Gannon, who had previously served for seven years in the very office that Rousselot now represented.[21] To carry out the assignment, the medical sciences division appointed a committee that was loaded with former military officials and educators whose institutions had benefited from military funding.* Gannon himself acted as staff for the committee.

---

*The committee was chaired by Norvin C. Kiefer, chief medical director of the Equitable Life Assurance Society of the United States, who had held jobs with the Defense Department and Civil Defense Administration before joining Equitable in 1953. Other members included Dewey F. Barich, president of the Detroit Institute of Technology, who had previously served as coordinator of veterans affairs for thirteen years at Kent State University; John C. Green, a lawyer who had recently left a high post with the Office of Emergency Planning; John M. Howard, a surgeon and professor at Hahnemann Medical College, which was a recipient of MEND funds; George V. LeRoy, medical director of Metropolitan Hospital in Detroit, who for more than a decade had served as associate dean at the University of Chicago, another recipient of MEND funds, and who had served as biomedical director for Operation Greenhouse, the first American test of a thermonu-

The result could have been predicted. In May 1969, the committee issued a report which called it "essential" that the MEND program be reinstated in fiscal 1970, with certain changes to improve its effectiveness.[22] And when it became apparent that the report would not be completed in time to support the Pentagon in House hearings on the 1970 budget, the committee obligingly dispatched a letter to Rousselot strongly endorsing "prompt reinstatement of funding."[23] On the basis of this letter, military officials told Congress that the Academy had concluded that MEND was "a good program, worthy of continuation."[24] But Congress was not impressed, and the Defense Department finally gave up the effort to restore funding. Although the Academy's report had no effect, the episode revealed how Academy units sometimes become willing accomplices of agencies with which Academy personnel have close ties.

A final control which the government exercises over the Academy is its ability to influence the wording of Academy reports. It is not generally realized that many of the supposedly "independent" Academy committees submit their reports to the sponsoring agencies for comment before making them public.

The dangers of the situation became apparent in another little-noticed incident in 1969 which put the Academy in opposition to the Bureau of Mines. This occurred when an Academy committee prepared a report, *Resources and Man*, which originally included a recommendation that the federal government expand its helium conservation program. The report stated that helium had unique qualities which made it highly valuable, if not essential, for such high-technology applications as superconductivity, supercooling, cooling of nuclear reactors, exploration of the seabed, and the space program.

This recommendation conflicted with plans under consideration within the Nixon Administration to abandon the program entirely on the grounds that enough helium had already been stockpiled to meet foreseeable needs for dec-

---

clear bomb in the early 1950s; Gordon E. McCallum, a former Assistant Surgeon General of the U.S. Public Health Service, which had provided financial support to MEND in its earlier years; and Steuart L. Pittman, a lawyer who had served as Assistant Secretary of Defense.

ades, particularly since new technologies might ultimately be developed that would allow helium to be extracted from the atmosphere or produced in fusion reactors. So the Academy recommendation was never released. Before the report was published, it was circulated to the sponsors who had provided financial support. One of those sponsors—the Bureau of Mines, which had been operating the conservation program—objected strenuously and claimed that it had more recent data, not considered by the committee, which raised doubts about the need for stockpiling further helium. Although the Bureau of Mines had no formal veto power over the report, the chairman of the Academy committee, Preston E. Cloud, Jr., a geologist at the University of California's Santa Barbara campus, yielded to the Bureau's entreaties and toned down the report so that it no longer called for expanding the conservation program but simply urged a reevaluation to determine whether the program should be extended.[25]

As it turned out, however, the Bureau of Mines didn't have any new information that would have changed the committee's mind—it simply used that claim as a ploy to defuse the Academy recommendation. "I think the director of the Bureau of Mines lied to me," Cloud said in a September 1972 interview. "There's not much you can do about that sort of thing."

Such cases suggest that the Academy is run, in some ways, by the government. For the Academy is not truly the master of its own destiny. The government largely determines what work the Academy will do, and in subtle—and unsubtle—ways it influences the course of the Academy's deliberations and the content of its reports.

The problem of direct industrial or commercial influence on the Academy is less extensive than the problem of governmental influence, largely because the government is by far the major sponsor of Academy studies. But certain units at the Academy have close ties to the business world and are susceptible to industrial biases. This is particularly true of the NAE, which operated as a largely autonomous sister unit of the NAS–NRC from 1964 to 1974, when its advisory operations were merged into the NRC.

The membership of the honorary NAE, which is supposedly elected primarily on the basis of creative engineering achievements, with leadership qualities a "supplementing" consideration,[26] is heavily weighted with top management. An NAS officer who made a study of the NAE membership as of early 1971 concluded that, at the time of their election, 43 percent were high-level corporate managers while another 22 percent were high-level executives in other fields. The British scientific journal *Nature* analyzed the Academy's description of the twenty-nine engineers elected to the NAE in April 1971 and noted that most were credited with management skills rather than with any specific engineering accomplishment. "If there existed an organization such as a national academy of successful business and government executives, its membership would probably not differ greatly from the National Academy of Engineering," *Nature* asserted. The journal found it "hard to understand how the National Academy of Engineering, with so many thousands of American engineers to choose from, can exhibit so little imagination as to draw its new membership almost exclusively from the top brass of the Defense Department, Ford, General Motors, General Dynamics, Lockheed and so forth."[27]

The NAS leadership was wondering the same thing. As Handler expressed it in a private letter to the NAS members in June 1971: "The Council—and many of our members—have been disturbed by the fact that, to an extent greater than had been anticipated, the individuals elected to the NAE have achieved recognition for their management skills in the world of technological industry rather than for their creative engineering; whereas, presumably, NAE was created to be an honorific body for distinguished practicing engineers."[28]

Clarence H. Linder, then NAE president, told the study team that a large proportion of the NAE membership comes from industry because that is where most engineering is practiced. He also justified the election of so many executives on the ground that a talented engineer typically reaches a managerial position at an early age. But justified or not, the corporate flavor at the NAE has raised serious questions of industrial bias in the NAE's past advisory work.

The leadership of the NAE is traditionally dominated by corporate executives. During 1974–1975 the president was

Robert C. Seamans, Jr., former Secretary of the Air Force; the vice-president was William E. Shoupp, senior vice-president at Westinghouse Electric Corporation; the treasurer was Edward N. Cole, president of General Motors Corporation; and the Council included, among others, Paul F. Chenea, a vice-president of General Motors, W. Kenneth Davis, a vice-president of Bechtel Power Corporation, John H. Dessauer, retired executive vice-president of Xerox Corporation, Edward L. Ginzton, chairman of Varian Associates, Robert C. Gunness, president of Standard Oil Company (Indiana), Frederic A. L. Holloway, a vice-president of Exxon Corporation, Clarence H. Linder, a retired vice-president of the General Electric Company, and Morris Tanenbaum, a vice-president of Western Electric.

Some of the NAE's advisory committees have been headed by high officials from corporations with a major stake in the subject matter considered by the committee. In 1972–1973, for example, the NAE's Aeronautics and Space Engineering Board, which advises the government on matters ranging from the space shuttle to civil aviation, was chaired by Willis M. Hawkins, a senior vice-president of Lockheed Aircraft Corp.; a committee that was advising on the design of research studies to assess the impact of the Bay Area Rapid Transit (BART) System in San Francisco was headed by Seymour W. Herwald, a vice-president for Westinghouse Electric Corporation, which holds a major subcontract on the BART system; the Marine Board, which advises on such matters as undersea technology and offshore resources development, was headed by William E. Shoupp, also a vice-president of Westinghouse, which has oceanic interests; and a newly formed Space Applications Board was headed by Allen E. Puckett, executive vice-president of Hughes Aircraft Company.

High NAE officials told us that the engineers appointed to NAE committees are expected to act as members of a professional academy rather than as representatives of their companies. They also said that, when they appoint a committee, they try to include enough diverse viewpoints so that no one man's bias will dominate. Thus, in constituting a Committee on Power Plant Siting, they placed Roland C. Clement, vice-president of the National Audubon Society, on its steering group, and a smattering of environmentalists on its various working panels. But the "public interest" representatives sometimes find the NAE atmosphere uncongenial. Two environmentalists appointed to the power plant siting study resigned in discouragement. Robert N. Rickles, commissioner of air resources for New York City, told us he could not accept the study's assumption that energy use was bound to increase. And Dean Abrahamson, director of environmental studies at the University of Minnesota, told us he dropped out because the procedures for drafting the report were unclear and "industry was pretty much calling the tune."

It would be misleading to suggest that every industrial executive acts as a lobbyist for his company's interests. But the participation of so many corporate executives has eroded the objectivity and independence of some NAE committees. As a result, some committees have served more as advocates or defenders of the industry position than as independent evaluators of national programs.

Almost all NAE committees start from the assumption that problems should be solved through the existing configuration of American industry rather than through approaches that would shake up the industrial status quo. A 1974 report, *U.S. Energy Prospects: An Engineering Viewpoint*, was perceptively criticized on these grounds by an article in *Science* which described the report as "an industry viewpoint on how to achieve energy independence." (A majority of the task force had industry ties.) As *Science* saw it:

> The report assumes, for example, no direct government involvement in the production of energy or the management of the energy industry. Instead, the task force urges the government to take prompt action to clear away red tape, provide incentives, and solve environmental problems, leaving industry free to get on with the job. In essence, the philosophy espoused is to push the existing energy system harder and to avoid creating any new institutional arrangements in the interests of getting things done quickly. Alternatives, such as wartime-like mobilization of the industry or expanded roles for federal agencies, are not considered.[29]

An NAE report that parroted the line of corporate interests even more blatantly was a 1970 study of the possibility of controlling sulfur oxide (SO_2) emissions from power plants and other stationary sources. The authoring committee was chaired by a retired DuPont Company technical director and most of its members were drawn from the ranks of industry. Not surprisingly, two of the report's major findings supported industry positions on pollution abatement.

The report's central conclusion was that, "contrary to widely held belief, commercially proven technology for control of sulfur oxides from combustion processes does not exist. Efforts to force the broad-scale installation of unproven processes would be unwise; the operating risks are too great to justify such action, and there is a real danger that such efforts would, in the end, delay effective SO_2 emission control."[30]

This finding was promptly used by industry as a shield to fend off a regulatory crackdown on SO_2 emissions. At a July 8, 1970, hearing in the state of Washington, for example, Charles F. Barber, board chairman of American Smelting and Refining Company, argued that one of his company's smelters should not be required to meet stiff emission standards because the "intractability of the problem" had been made "abundantly clear" in the Academy report.[31] Similar claims, based on the NAE report, were made in October 1970 by a spokesman for Union Carbide Corporation[32] and in June 1970 by the general manager of the Utah Copper Division of Kennecott Copper Corporation.[33] Such industry maneuvers led William Megonnell, a top enforcement official with the National Air Pollution Control Administration (NAPCA), and its successor agency, the Environmental Protection Agency, to complain in a 1971 interview that the Academy's sulfur oxide report had had "an adverse impact on my operation." He explained: "Every time I went to an abatement conference and the subject was sulfur oxide controls, this Academy report was flashed in front of me to show that there was no commercially proven technology."

There seems to be general agreement that the Academy was right in asserting that no technology was yet "commercially proven" in full-scale plant operations at the time the report was issued in early 1970. But there is considerable disagreement as to whether that finding justified the conclusion that "efforts to force the broad-scale installation of unproven processes would be unwise." Officials of NAPCA took an opposing view—namely, that the technology had been proved in pilot plant operations and should therefore be used on a commercial basis to protect the public health and welfare. At a November 1970 abatement conference that dealt with the Union Carbide plant in Marietta, Megonnell took issue with the terminology used by the Academy report. "The words they use—it's not commercially proved. That is quite different in NAPCA's opinion from commercially feasible. We think the technology is well proved, and now it is a matter of scaling up commercially to do the job. It's a fine distinction perhaps, but an important one. It never will be commercially proved until somebody is forced to prove it."[34]

The decision on whether to require a substantial reduction in sulfur oxide emissions is a complex one involving a careful weighing of the costs and benefits in terms of public health, environmental quality, impact on the economy, and related issues. None of these issues was given more than fleeting consideration by the NAE committee. Indeed, the committee, which consisted primarily of engineers, was no more qualified than any other group of citizens to judge what would be "wise" public policy in light of all the factors involved. Its conclusion must be read as an expression of the prevailing attitude in industrial circles.

The NAE report went on to recommend "a high level of government support" for research, development, and demonstration of the more promising technologies, but it did not call for a similar investment by the power companies. Indeed, the report suggested that utilities cannot afford to spend much money on pollution research because the various bodies that regulate utilities don't allow adequate reimbursement for such expenditures. "Funds spent by utilities for development and application of pollution control processes may not be readily included in their capital structure, which is the basis for establishing consumer rates," the report said.[35] But the implication that utilities can't support extensive research of their own was sharply disputed by environmentalists. Senator Lee Metcalf (D-Mont.) pointed out that utilities were then allowed to count R&D costs as operating expenses, a circumstance which made it possible to pass such costs on to the consumers. If such costs were included in the capital structure, he added, then the utilities would actually make a profit on them.[36] The issue is a complex one which we need not explore. For our purposes, it is enough to note that the NAE committee was expert in engineering and technology, not in the intricacies of utility financing and regulation. Yet it felt no qualms about uncritically parroting the power industry's explanation for its failure to do more research.

Corporate influence has also served to restrict the kinds of projects the NAE is willing to undertake. Indeed, corporate officials are sometimes in a position to head off studies or reports that might prove troublesome to their industries. An example of this occurred in 1971 when an executive of the Standard Oil Company of New Jersey (now called Exxon) played a major role in the non-publication of a pollution tax study that had been prepared by the staff of the Committee on Public Engineering Policy (COPEP). That report, entitled *Strategies for Pollution Abatement*, compared the use of direct regulation, economic incentives, and litigation as techniques for controlling pollution. Its central conclusion was that "emission charges can be a powerful policy implement in the abatement of pollution" and therefore "ought to be given a try." The report also suggested that power plants, refineries and other emitters of sulfur dioxide should be the first target for such emission charges.[37]

The report was reviewed by an ad hoc group of four COPEP members. Three of the reviewers—Charles J. Meyers, a Stanford University law professor; Joseph Fisher, an economist who heads Resources for the Future, Inc., a Washington-based think tank; and Ruben F. Mettler, president of TRW, Incorporated—felt the report was sufficiently promising and important that it should be revised to eliminate weak spots and then issued as a COPEP document. But the fourth member of the review panel was vehemently opposed to publication. He was Frederic A. L. Holloway, vice-president for corporate planning at Exxon, a company that would be adversely affected by the report's recommendations. Holloway told us that he was not trying to protect Exxon from having to clean up its discharges; nor was he opposing pollution abatement per se. Rather he felt that the report's advocacy of emission charges was "too simplistic" and that regulatory standards would be more effective in controlling pollution than would a pollution tax. "I didn't think the study was good enough," he explained. "I wasn't trying to quash something that would do a better job of pollution control." Staff members assert that Holloway also submitted a critical memorandum from an Exxon economist, though Holloway said he does not recall using his corporate staff to review the report.

The report was discussed at two meetings of COPEP in early 1971. Holloway was joined in his opposition by several other COPEP members, with the result that the committee, at its March 15, 1971, meeting, decided not to publish the report under the NAE imprimatur but agreed to allow the authors to arrange publication on their own. Thus the NAE as an institution had been blocked from participation in the national debate then shaping up over a pollution tax on sulfur. On February 8, 1971, just as COPEP was reviewing its staff report, President Nixon submitted an environmental message which suggested that "a charge on sulfur emitted into the atmosphere would be a major step in applying the principle that the cost of pollution would be included in the price of the product."[38] A year later, the Nixon Administration and various Congressmen introduced legislation embodying alternative sulfur tax plans. Support for the tax approach was voiced by the Council of Economic Advisors, the Council on Environmental Quality, Nobel Prize–winning economist Paul A. Samuelson, conservative economist Milton Friedman, and numerous other economists and environmental groups. Virtually unanimous opposition was expressed by the oil, coal, power, and smelting industries.[39]

The staff members of COPEP, acting as individuals, made their views known and even lobbied for a sulfur tax. But neither COPEP nor the NAE made any contribution to the debate. They had been effectively removed from the arena by a small group of COPEP members, of whom the most outspoken was an industrialist whose corporation would have been adversely affected.

The NAE has, on occasion, served as a vehicle through which industry leaders and government officials with common interests collaborate. The result is that the NAE loses all pretense of independence and simply becomes an adjunct to a government program. That seems to have happened to the NAE's Aeronautics and Space Engineering Board (ASEB) which has been working closely—and confidentially—with the National Aeronautics and Space Administration (NASA) on such controversial projects as the space shuttle and the development of advanced supersonic technology.

The space shuttle—which would fly crews and cargo into orbit, then return to earth largely intact, ready for another mission—is NASA's biggest development effort for the 1970s. It is strongly endorsed by the ailing aerospace industry, which seeks to recover from a substantial loss of federal contracts in the early 1970s. The justification for the shuttle is that it will eventually reduce the cost of space missions because the shuttle would be reusable, unlike the current expendable boosters. But the shuttle project has been attacked by economy-minded senators, and by individual scientists and scientific groups as well, on the grounds that it will not be economic at the volume of space traffic anticipated, that it makes little sense to develop a costly new transportation system when reliable expendable boosters already exist, and that the funds poured into developing the shuttle will detract from support of other important NASA activities.

The shuttle issue is probably the most important debate involving NASA programs in this decade. Yet who did the NAE choose to provide "independent" advice on the program? None other than a board dominated by the very aerospace interests which are pushing for a shuttle. The ASEB was headed in 1973 by two aerospace officials—Willis M. Hawkins, the Lockheed vice-president, as chairman; and George E. Solomon, vice-president and general manager of TRW Systems Group, as vice-chairman. Its other members included retired NASA officials, aeronautics engineers from universities, and industrial executives from a variety of firms with interests in aerospace projects, while its staff director was a former Air Force officer.

The board has typically acted as a private technical consultant for NASA. It has listened to presentations by NASA technical officials, asked questions or raised objections based on the experience and knowledge of its members rather than on a detailed study, then briefed the highest-ranking NASA administrators orally and in a privileged follow-up letter that is not available for public inspection. The justification for this procedure, according to Hawkins, is that it has enabled the board to respond promptly to NASA's need for advice without taking the time to develop a polished report and submit it for review by other experts within the Academy. But the result of this mode of operation is that the NAE has abandoned the role of objective critic. The NAE has made no effort to conduct an independent evaluation of whether the space shuttle or the supersonic transport should be built. Instead, it has designated a committee of aerospace advocates to serve as a private consultant to the space agency, helping the agency to refine its plans, improve the proposed configuration for the shuttle, and modify the cost and pace of the development program. But it has never raised questions as to the basic worth of the programs.

The industry-biased performance of some NAE committees may be mitigated by the 1974 merger of the NAE's advisory operations with those of the NAS–NRC. As a result of the merger, appointments to engineering committees will henceforth be subject to the more formalized conflict-of-interest screening in effect at the NAS, and reports issued by engineering committees will have to clear the report review hurdles at the NAS. Both factors may tone down the industry voice in engineering reports. However, the corporate executives who control NAE will be given control of the Academy's Assembly of Engineering, the NRC unit into which most NAE committees will be folded. And NAE members will be appointed to the Academy's report review apparatus. Thus it is possible that the corporate influence which has dominated the NAE may begin to exert greater influence over the reports issued by the Academy itself.

Top NAE officials repeatedly told the study team that industrial bias is more insidious at the NAS, with its university orientation, than at the NAE, with its corporate ties. Their point was this: If a man is a corporate vice-president, he wears that title, and the public is immediately aware that he may have a corporate interest in the outcome of an issue. But if a professor working for the NAS has close ties with industry, perhaps through a consulting arrangement or a research grant, this is seldom well known, for he generally wears only his academic title.

The industrial voice enters NAS deliberations in many ways. In some cases, industry scientists serve as members of Academy committees. Thus, a DuPont scientist was allowed to draft key sections of a report on lead pollution; a group of industrial toxicologists was allowed to draft guidelines for determining insignificant levels of chemicals in food; a sub-committee on dog and cat food standards was headed from 1968 to 1973 by an official of Ralston Purina Co., a major pet food manufacturer; and the Maritime Transportation Research Board was headed in the early 1970s by the president of Luckenbach Steamship Co., Inc. About 17 percent of the individuals who sat on NRC committees in 1970 worked directly for industry or were self-employed.[40]

In addition to such direct industry representation, many academic scientists serving on Academy committees have ties to industry that are difficult for an outsider to detect. Thus a committee that issued a report in 1971 on the biological effects of airborne fluorides was composed entirely of scientists from universities and research laboratories that were seemingly independent of industry influence. But it was later revealed that four of these scientists, who had written most of the report, had close ties to the aluminum industry, which is a major emitter of fluorides. Some had written publications for the Aluminum Association, received research support from the industry, or testified for the industry in hearings on fluoride standards.[41] The report which they helped prepare under the Academy imprimatur proposed tolerance thresholds which were somewhat more lenient than standards proposed by the Center for Science in the Public Interest, a Washington-based study group.[42]

Even when industrial scientists are not members of a committee, and are not publicly listed as contributors to a report, they are sometimes able to influence a committee's recommendations by behind-the-scenes lobbying. This became apparent in 1971 when the Academy's Committee on Food Standards and Fortification Policy submitted a report to the Food and Drug Administration recommending nutritional guidelines for TV dinners. The chief manufacturers of TV dinners gained a voice in the committee's deliberations through a side door. The committee appointed five "ad hoc advisers" to assist in the study, of whom two were top officials of the major TV dinner makers, namely Robert H. Cotton, a vice-president of ITT Continental Baking Co., and Carl Krieger, president of the Campbell Institute for Food Research. Meanwhile, a consumer champion who sought a voice in the study—Robert Choate, well-known for his crusade to improve the nutritional quality of breakfast cereals—says he requested an opportunity to participate but was turned down.

The influence of the industry "advisers" was exerted shortly after the committee submitted its first progress report to the FDA. In the original progress report, which was treated as "confidential," the committee recommended minimum levels for several specific nutrients, including a recommendation that each TV dinner should supply 50 percent of the calcium an individual is supposed to get each day in accord with recommended dietary allowances. The committee took this stand on nutritional grounds, arguing that, unless the amount of calcium in TV dinners was boosted, "there will be a nutritionally undesirable calcium-to-phosphorus ratio, which is difficult to correct by supplementation of the diet with dairy products alone."[43] But, shortly after the report was delivered to FDA, it was shown to the ad hoc advisers at a closed meeting where it evoked sharp criticism. According to one participant in the meeting, the two representatives of the TV dinner manufacturers "took over the meeting for thirty minutes or so" to complain at length that existing food technology did not permit them to add more calcium to frozen dinners without lowering the taste and texture qualities. There was no rebuttal to the industry contentions, and, when the committee submitted its final report on TV dinners, it excluded calcium as "not appropriate for frozen convenience dinners." The report explained that "because of technological difficulties, fortification of this type of product with even small amounts of calcium does not appear feasible at this time."[44] It did not explain how the committee had reached this conclusion, nor what documentation it could offer to substantiate its assertion. The FDA adopted, with minor changes, all the nutrient levels recommended by the Academy, including the recommendation that no minimum level be set for calcium. Thus a national policy was set on the basis of what two industry representatives, arguing behind the closed doors of the Academy, claimed was technologically feasible. Whether the technological difficulties were really as great as claimed is difficult to ascertain, for there was no strong consumer voice on the committee to challenge the contention of the industry.

The Academy's recommendations were greeted with dismay by consumer groups. The Academy had recommended minimum levels for six different nutrients, but the levels recommended were lower than the levels currently contained in most TV dinners. Of thirty-five commercial TV dinners surveyed by the Academy, for example, only eight contained less protein than the level recommended by the Academy, and only five contained less iron.[45] Robert Choate's Council on Children, Media, and Merchandising complained that the proposed guidelines were "a model for legislating the *status quo*" and were "more reflective of industry perspectives than of nutritional common sense."[46] Similarly, Public Citizens' Health Research Group, affiliated with Ralph Nader, charged that the guidelines represented "the lowest common denominator approach to setting food standards."[47] If manufacturers choose to do so, they can actually reduce the nutritional quality of most of their dinners and still exceed the Academy's recommended standards.

The financing of some Academy studies also raises questions as to their objectivity. Some Academy units, such as the Building Research Advisory Board, the Food and Nutrition Board, the Drug Research Board, the Agricultural Board, and the Computer Science and Engineering Board (which was abolished in 1972) receive financial support from industry. Academy officials say they never perform contract work for a single company, but they do accept funding from companies to help support work that is of interest to a broad segment of industry. These contributing companies are then in a position to make suggestions as to what studies should be carried out and how they will be carried out. Moreover, the Academy has, in the recent past, performed work for industry-financed organizations that have a stake in a favorable verdict. A prime example of this was a 1969 study of roadside litter conducted by the Academy's Highway Research Board to obtain "unbiased information regarding the nature and composition of litter."[48] The study was requested and financed by Keep America Beautiful, Incorporated, an organization that was established in 1953 by bottle and can manufacturers and other companies whose products contribute to the litter problem. KAB is strongly opposed to laws aimed at the manufacturers; instead it pushes for a crackdown on the little guys who litter the roadside. KAB has a three-point answer to the litter problem: educate the public not to drop litter, punish those who do drop it by strict enforcement of antilittering laws, provide more disposal facilities.[49] Although KAB clearly has a strong vested interest in the issue, the Academy apparently had no qualms about undertaking the project. It even allowed three KAB representatives to maintain close liaison with the design and development of the project.[50]

The findings of the study could not have been more pleasing to KAB. The main thrust of the study, which was coordinated for the Academy by researchers at the Research Triangle Institute in North Carolina, involved collecting litter from sections of highway in twenty-nine states and counting the individual items. The count showed that only 22 percent of the litter items were cans, bottles, or jars, while 59 percent were paper.[51] The Highway Research Board, in a foreword to the study report, drew the conclusion that the bottle and can manufacturers were looking for: "Control of cans and bottles alone will not solve the problem.... It would seem that publicity efforts led by Keep America Beautiful Inc. and the state highway departments continue to offer the most practical approach to the problem of reducing litter."[52]

This verdict was so pleasing to the manufacturers that they began using the Academy report itself in their "educational" campaigns. In 1970, when the manufacturers were threatened by a proposal in Washington State which sought to eliminate throwaway containers by requiring a five-cent deposit on all containers for beer and soft drinks, the industry forces, in alliance with labor unions concerned about loss of jobs, launched a massive publicity campaign that succeeded in defeating the measure.[53] One of their main arguments was that the five-cent deposit was unnecessary because the Academy report had shown that cans and bottles were only a small part of the litter problem.[54]

Proponents of the measure retorted that the Academy report was defective and biased. William H. Rodgers, Jr., a law professor who had helped draft the deposit proposal, later called the Academy's method of counting individual items of litter an "absurdity" because it managed "to equate a bottle or can with a scrap of newspaper or a dead porcupine," thus ignoring such matters as volume, weight, and long-term threat to the environment. "When you count items and ignore such questions as their disposability, weight, the environmental impact, and so on, you have got skewed data," he said. "Other studies have indicated ... that the bottles and cans, instead of being a minor percentage of the litter problem, may well be a significant percentage."[55] A survey by the Oregon State Highway Division, for example, concluded that, on a volume basis, bottles and cans constituted 62 percent of the litter problem—not 22 percent as the Academy claimed.[56]

An analysis published in 1973 by the Environmental Protection Agency concluded that an Oregon law which required deposits on beverage containers brought about "a significant and positive impact on litter in Oregon."[57] Thus, whatever justification the Academy's count-the-item approach might have, there is little doubt that it provided scant basis for the sweeping conclusion that the litter problem would best be solved through "publicity efforts" led by Keep America Beautiful, Incorporated. But should one expect otherwise from a study commissioned and financed by the very industries causing the problem under investigation?

A pernicious consequence of the industrial presence on some Academy committees is that it inhibits the committees from venturing into areas that might be viewed as dangerous to powerful segments of industry. Such a complaint was lodged against the Drug Research Board by James L. Goddard, former commissioner of the FDA. Writing in *Esquire* in 1969, Goddard described the board as part of "The Drug Establishment," which he defined as "a close-knit, self-perpetuating power structure consisting of drug manufacturers, government agencies and select members of the medical profession." Although Goddard acknowledged that the board represented "the best of government, industry and science," he added: "it is, by its very membership, a limited forum in which consensus is developed for the Establishment." His most specific complaint was that when he had advocated the publication of a national drug compendium that would provide physicians with a single source of information on all drugs currently on the market—not only the brand names marketed by the big companies, but also the cheaper equivalents put out by lesser-known companies—the Drug Research Board was reluctant to endorse the idea. "When the Compendium was first discussed with the NAS–NRC Drug Research Board, it met with a degree of acceptance," Goddard wrote, "but not enough to overcome the Establishment's opposition to having its products listed on an equal basis with those of non-Establishment companies. The lack of the Board's unqualified support was a factor in the failure of Congress to act on a bill to establish the Compendium."[58]

The Building Research Advisory Board (BRAB), one of the Academy's largest subunits, has similarly been wary of issues that might disrupt the status quo in the building industry. The board has traditionally been dominated by industry representatives and has received part of its funding from industry. In the opinion of George B. Kistiakowsky, former Academy vice-president, it is "probably an industrial lobby par excellence operating within the Academy structure." The overt manifestation of this, Kistiakowsky says, is that "BRAB never had the guts or the urge to challenge the building industry or the local political structures about archaic building codes, about lack of quality in construction, or about the absurd cost of construction caused by interaction between the industry and the unions preventing automation of construction." Concerned about the industrial ties of BRAB, the Academy in the 1970s began to broaden the composition of the board and made plans to lessen the input of industry groups.

Another instance of an industry-inhibited committee is the Highway Research Board's Committee on Vehicle Characteristics, which is supposed to consider "vehicle characteristics and vehicle performance factors related to safe, efficient, and economical operation of highway facilities."[59] That broad mandate seemingly embraces virtually all aspects of vehicle safety. But the committee has actually done little or no work on vehicle safety. Instead, it has played a lethargic liaison role between the automobile manufacturers and the highway designers.

What accounts for the committee's passivity and its reluctance to get into the touchy issue of the safety of the automobile itself? Part of the explanation can be found in the committee's leadership. Chairman William A. McConnell is a top engineer for Ford Motor Company, while the committee secretary, Kenneth Stonex, is from General Motors. With the top two automobile companies in the driver's seat, there is slim likelihood that the committee will do anything that they find threatening.

The most obvious special interest operating within the Academy complex is the scientific community itself. Although the Academy's major purpose, as expressed in the charter, is to advise the government, it has traditionally pursued a second major objective as well: protection and promotion of the interests of the American scientific community. When the Academy is called upon to advise the government on issues that might affect the fortunes of the scientific community, it generally behaves more like a self-serving advocate than a disinterested adviser.

The Academy's desire to nourish the scientific community is perhaps the strongest drive operating within the institution. It can even overcome the Academy's tendency to remain docile and subservient in its relationships with the government. We repeatedly asked Academy officials to cite cases where the Academy had come into open conflict with government policy. The number of cases they cited was small, and almost every one involved a matter where the Academy felt the government was undermining or neglecting the prerogatives of the scientific community. Government actions which have sparked criticism from the Academy over the years include the firing of a government science administrator by political leaders who disagreed with the administrator's findings on the efficiency of a particular battery additive; security restrictions and loyalty oaths imposed on scientists; the Nixon Administration's decision to cut support of graduate students; and the Agriculture Department's alleged failure to support basic research adequately—all matters which were perceived as harassment or neglect of the scientific community.

The Academy often makes no pretense of appointing objective committees when the interests of the scientific community are at stake. Late in 1973, for example, the Academy appointed a committee to study the implications of a White House decision to abolish the science advisory apparatus which had served the presidency for several administrations. The committee was overwhelmingly dominated by men who had served in the old advisory mechanism. It was headed by a former presidential science adviser, eight of its other twelve members had served on the former President's Science Advisory Committee, and its staff officer was a former executive at the now-defunct White House Office of Science and Technology.* Thus it was no surprise when this group concluded in a 1974 report that the White House needed a new Council for Science and Technology to replace the apparatus that had been dismantled.[60] Whatever the merits of its recommendations, the report must be read as a brief by the scientists who had been thrown "out" as to why they should be allowed back "in" at the highest policy-making levels.

The most ambitious Academy effort on behalf of the scientific community over the past decade has been a series of disciplinary studies sponsored by the Committee on Science and Public Policy (COSPUP). These surveys, covering such fields as astronomy, physics, chemistry, biology, and mathematics, attempt to assess the state-of-the-art in a specific discipline and the extent of public funding that is needed to allow proper development of the discipline. The tone of these reports is unabashedly promotional—they invariably request substantial increases in funding for whatever discipline is being surveyed. Even some of the Academicians have become embarrassed by the tone of special pleading. Garrett Birkhoff, a Harvard mathematician who is a member of the Academy, has complained that the NAS and NRC serve as "sounding boards for 'special interest' groups advocating more funds to support their special interests, whether these be high-energy accelerators, seismology, mathematics, or more money for our own universities."[61]

---

*The committee was headed by James R. Killian, Jr., former science adviser to President Dwight D. Eisenhower. The eight committee members who had served with the old President's Science Advisory Committee included Ivan L. Bennett, Jr., Harold Brown, James B. Fisk, Edwin H. Land, Franklin A. Long, Emanuel R. Piore, Kenneth S. Pitzer, and Charles H. Townes. The staff executive assistant was David Beckler, a long-term official of the old Office of Science and Technology. The other four committeemen were Graham T. Allison, Jr., a young public policy specialist from Harvard; Robert C. Gunness, vice-chairman of Standard Oil Company (Indiana); Donald B. Rice, president of the RAND Corporation; and James Tobin, Yale University economist.

And Philip Abelson, a former member of the Academy Council, described the reports issued up to 1972 this way in an editorial in *Science:* "Since the reports have been so obviously self-serving, it is not surprising that the prodigious efforts devoted to them have come to little.... The procedure of asking representatives of a discipline to prepare material on their own field has some merit. But the experience of many years and many reports bears out the bromide of not asking the fox to guard the henhouse."[62]

The Academy seems to be trying to make its disciplinary reports more objective. We were told by members of some recent survey groups that they had been warned to submerge their promotional tendencies. And Handler has told his members that "the Academy must exhibit a higher order of statesmanship than does a self-serving lobby."[63] But such warnings serve merely to eliminate the most blatant excesses. The fundamental problem remains—the Academy is trying to be an advocate for the scientific community, and an advocate can seldom be objective.

The Academy's promotional instincts sometimes lead it to omit data that could imperil the financial fortunes of the scientific community. An example of this is in one of the Academy's most widely praised disciplinary surveys, the 1965 report entitled *Chemistry: Opportunities and Needs.* That report called for a rapid increase in federal support of basic research in university chemistry departments—from a level of about $50 million in fiscal 1964 to a level of about $120 million in fiscal 1968. One of the reasons given was that "Our survey has led us to conclude that research opportunities and manpower needs in chemistry warrant a more rapid growth in funding."[64] But the report was remarkably vague when it got around to discussing the manpower situation in terms of projected supply and demand for chemists. The committee acknowledged that its recommendations for greater support of chemistry would "affect the number and quality of chemists." But it added that manpower assessments "depend partly upon economics rather than chemistry, and therefore lie partly outside our field of special competence." The committee concluded that there should be "further joint study by economists and chemists."[65]

But that was a deliberate obfuscation. As Frank H. Westheimer, the Harvard chemist who headed that study acknowledged in an interview with the study team: "We knew damn well that the demand for chemists wouldn't last. . . . We never quite said this—but we also resisted the temptation to come out and say we had to train lots and lots of chemists. Instead, we urged a joint study by chemists and economists. It was not our proudest moment, but we didn't fall flat on our face." As it turned out, that joint study was never undertaken, and five years later an American Chemical Society survey estimated that 5000 chemists and chemical engineers were out of work while another 12,000 were in varying states of insecurity ranging from temporary unemployment to part-time employment to postdoctoral positions.[66]

A similar reluctance to issue information that might prove damaging to the financial support of science may also have influenced an Academy decision in the early 1960s against publishing a detailed study of the financing of graduate student education. The Academy's Office of Scientific Personnel, which helps federal agencies distribute fellowship awards to science students, asked the National Opinion Research Center (NORC) at the University of Chicago to design and execute the study; it arranged for Ford Foundation funding; and it supplied the chairman for an advisory committee to oversee the project.

The Academy has long advocated substantial federal funding for the graduate education of scientists. But the NORC study reached conclusions that undercut the need for such funding. The study found that "financial worries are low among American graduate students" and that "among advanced natural science students, except possibly in lower-stratum private schools, stipend levels are so high [with some 85 percent of the students receiving assistantships or fellowships] as to have approached something like a saturation point." The study also undercut traditional arguments that fellowships support outstanding but needy students when it concluded that "the chances of receiving a stipend depend more on 'where you are' than on what you need or how good you are academically."[67]

After reviewing the document, the Academy-led advisory committee refused to authorize its release as either an Academy or a committee document because such publication "would imply a degree of endorsement."[68] Peter H.

Rossi, who headed NORC at the time, charges that "They refused to publish the report because they didn't like the conclusions—they brought in all kinds of experts to find technical defects, and, since no study like this is ever devoid of defects, they found them." Similarly, James A. Davis, the NORC researcher who directed the study, told us: "The NAS is in the business of dealing out fellowships. . . . There was jolly little enthusiasm for the report."

This recital of the role of special interests in Academy deliberations should not be understood to imply that all Academy study groups fall prey to such influences. One can find instances where Academy committees, dealing with important public issues, have shown substantial independence from government and industry viewpoints. Moreover, it should also be acknowledged that the Academy tries harder than most consulting groups to eliminate bias from its advisory committees. In August 1971 the Academy imposed a requirement that appointees to its most sensitive committees must submit a statement listing "potential sources of bias." The initial version of this statement simply asked each prospective committee member to identify any current or previous affiliations, activities, financial interests, or public statements which others might construe as compromising his objectivity.[69] But a later version, put into effect in January 1973, was more explicit in its probing. It asked each prospective committee member to list all jobs, consultantships, and directorships held for the past ten years; all current financial interests whose market value exceeded $10,000 or 10 percent of the individual's holdings; all sources of research support for the past five years; and any other information, such as public stands on an issue, which "might appear to other reasonable individuals as compromising of your independence of judgment."[70]

The bias statement is used in the appointment process for all committees that are deemed "bias sensitive" by the Academy president's office. In mid-1973 about one-third of all appointees to Academy committees were being asked to submit bias statements, and only an occasional holdout was refusing to comply, according to S. Douglas Cornell, the assistant to the NAS president who is responsible for reviewing committee appointments. Meanwhile, every

Academy committee, whether exempted from the bias statement or not, is directed to discuss potential bias at least once a year.

The Academy's approach has sparked considerable resentment in some segments of the technical community. After the initial version of the bias statement was put into effect, for example, the Society of Automotive Engineers (SAE) complained that the Academy was showing "distrust of, and lack of faith in," professional scientists and engineers. The SAE boasted that it had taken "a different posture"—one which assumed that "engineers are mature and professional enough to separate out their prejudices and biases and to face objectively a problem before an SAE committee."[71] But the Academy's approach is surely the more realistic. It recognizes that any scientist, however mature and professional he might be, will approach a problem with value judgments that reflect his experience and his professional and economic interests. Thus, to achieve either a neutral or a balanced committee, it is crucially important to know what experiences the committee members have had.

The bias statement is a significant step toward enhancing the objectivity of Academy reports. But it is not a sufficient step. For one thing, the Academy refuses to make the bias statements public; thus it is difficult for outsiders to form an independent judgment as to whether any particular committee is impartial or biased. For another thing, the bias statement does not eliminate the structural, financial, legal, and other constraints that undermine the Academy's independence and objectivity. We shall propose our own solutions for these perplexing problems in our final chapter. Meanwhile, let us examine closely how the Academy has handled some of the major technological issues of our time, for it is only through such detailed scrutiny that one can appreciate how our most prestigious scientific institution so often fails the public interest.

CHAPTER FIVE

Radioactive Waste Disposal:

The Atomic Energy Commission Brings the Academy to Heel

"The Academy is a valuable asset to the nation; its success in its advisory role to government rests on a fragile and complex web of confidence: public confidence in the integrity and independence of the scientists who serve without pay on its many committees; agency confidence that they will receive usable advice given in a spirit of helpfulness, and that they will not be pilloried for either real or fancied wrongs. If the one confidence waned, the [Academy's] ... value to the nation would be severely depleted; if the other confidence were eroded, the Academy would indeed be out of business, for no agency would contract with it for advice. If one knows this operational frame, one understands why the Academy tends to be circumspect in its dealings with sponsoring agencies and close-mouthed about them afterward."

—Earl Cook, former Academy staff officer, explaining why the Academy allowed the AEC to suppress one of its reports and fire one of its committees, in a letter to U.S. Senator Frank Church (D-Idaho), May 1, 1970.

The Academy's efforts to advise the Atomic Energy Commission (AEC) on disposal of radioactive waste materials reveals how difficult it is for the Academy to function as an effective critic of government. Since the mid-1950s, the Academy has been advising the AEC on the possibility of burying wastes deep beneath the earth in stable geologic formations. For the first decade of this effort, the Academy adopted an aggressive, independent stance. It suggested avenues of investigation which profoundly shaped the AEC's approach to the problem, and it was highly critical of existing waste management operations at AEC installations.

But the critical views were buried by the AEC, with Academy acquiescence, and the committee which authored them was disbanded at AEC insistence. It was replaced by a new committee closer in philosophy and background to the AEC's own thinking. Although the new committee has not been a mere rubber stamp for the AEC, its reports have tended to be supportive of AEC plans. Opinions differ as to whether the new committee or the old committee is more sound in its judgments. But whatever the merits of their respective positions, the circumstances surrounding this particular Academy advisory effort suggest that the Academy has identified too closely with the interests of the sponsoring agency—to the detriment of independence in behalf of the public interest.

Radioactive wastes present one of the major unsolved problems facing the nuclear age.[1] Such wastes are generated wherever radioactive materials are used. The mining, milling, and preparation of uranium fuel for reactors and weapons, the irradiation of nuclear fuels in reactors, the chemical processing of irradiated fuel to recover usable uranium and plutonium—all these and other steps in the handling of radioactive materials produce wastes that are extremely toxic. The great bulk of the wastes produced so far has been generated by the AEC itself, mainly as a byproduct of weapons manufacture. But, as civilian nuclear power plants continue to proliferate, commercial wastes will become increasingly troublesome. The wastes range in toxicity from the so-called "low-level wastes," which are released to the environment after dilution or simple processing, through "intermediate-level wastes," to the "high-level wastes," which pose the greatest health hazard and the most complex technical problems of management. The only practical way to reduce the radioactivity in these wastes to nonhazardous levels is to allow the radioisotopes to decay naturally, a process which takes hundreds of years in the case of such toxic substances as strontium and cesium and perhaps 200,000 or more years for plutonium 239. Thus a secure means must be found to keep these wastes isolated from the environment, and particularly from the food and water supply, for centuries.

The wastes generated so far—some 80 million gallons as of 1971—have been stored in tanks or other holding facilities on an interim basis. In theory, such storage in adequately designed tanks could protect the public indefinitely, but the tanks deteriorate in a matter of decades, which means they require extensive surveillance, continuing maintenance, and perpetual replacement—all for periods longer than the life expectancy of governments. Thus the AEC—which is responsible for managing its own wastes and for regulating the wastes generated by the commercial nuclear industry— has been seeking some spot in the universe where the wastes can be placed and more or less forgotten. Alternatives that have been suggested range from shooting the wastes into outer space (a costly procedure which involves the hazard that a rocket might go astray and dump the wastes right back on earth) to burying the wastes in various geologic strata deep beneath the earth. So far, no final solution has been found. The permanent disposal of radioactive wastes remains one of the "most troublesome problems" confronting the nuclear enterprise, according to Alvin M. Weinberg, director of the AEC's Oak Ridge National Laboratory and a member of the NAS as well.[2]

The Academy was first asked in 1954 to help the AEC evaluate methods of burying the high-level wastes underground. A steering committee was formed which sponsored a major conference of experts at Princeton University in

September 1955; and a more permanent committee was set up in late 1955. The committee had a geologic focus and remained part of the Academy's earth sciences division for more than a decade. Although its membership changed gradually over the years, and its name changed as well—it was the Committee on Waste Disposal from 1955 to 1960 and the Committee on Geologic Aspects of Radioactive Waste Disposal thereafter—the group was essentially a single continuing committee. It repeatedly argued that the AEC was pursuing expedients that might jeopardize the safety of future generations, and it deliberately adopted a critical stance aimed at goading the AEC into adopting what the committee regarded as sound disposal practices.

The committee's first major report, published in 1957, warned that "The hazard related to radioactive waste is so great that no element of doubt should be allowed to exist regarding safety." It recommended that the wastes should not come into contact with any living thing during their period of toxicity, which might be six hundred years or more. And it concluded that the wastes could safely be stored in stable geologic formations beneath the earth. The committee particularly favored disposal in caverns mined in salt formations, preferably after the liquid wastes had been converted to a more immobile solid form.[3]

The suggestion that the wastes be buried in salt formations was a major contribution by the Academy toward resolution of the problem. Salt has many advantages as a disposal medium—it flows plastically, thus healing any fractures which might develop and effectively sealing in the wastes once they are buried; it is a good radiation shield; it dissipates heat better than other types of rock; it is found in areas that are free of earthquake hazards; it has almost always been geologically stable for millions of years; and its very existence is evidence that it has not been in contact with water or else it would have dissolved.[4] Burial in salt seems to be the favored long-term solution in both the United States and West Germany, and the Academy deserves major credit for focusing attention on salt's desirable features. The Academy committee functioned creatively in this instance. It did not merely review AEC plans and pick the best alternative; it actually produced a new idea that pushed AEC programs in a new direction. Even the

AEC acknowledges that the Academy "first suggested the use of salt formations."[5]

The committee and the AEC soon came into conflict, however, most notably over the choice of a location for permanent waste storage. The committee wanted the AEC to use the best possible geologic structures, but the AEC, for budgetary and convenience reasons, wanted a disposal site at each of the major AEC plants where high-level radioactive wastes are generated—the Hanford Plant in the state of Washington, where plutonium for nuclear weapons is produced; the Savannah River Plant in South Carolina, also a weapons facility; and the National Reactor Testing Station in Idaho, where nuclear reactors are built and tested. "They pressured us right from the start that they wanted a disposal site at each of these plants," recalls M. King Hubbert, a geophysicist who was associated with the committee for most of its existence. "They never let up on this. They kept harassing us."

In 1960, the committee took the unusual step of voicing its concerns in a letter to the AEC commissioners rather than to the Division of Reactor Development and Technology, the subunit of the AEC which the committee was officially advising. Summing up its conclusions after five years of advisory work, the committee said:

"No existing AEC installation which generates either high-level or intermediate-level wastes appears to have a satisfactory geological location for the safe local disposal of such waste products; neither does any of the present waste-disposal practices that have come to the attention of the Committee satisfy its criterion for safe disposal of such wastes."[6]

The committee recommended that "urgent" action be taken to establish facilities at suitable geologic sites where the wastes that had accumulated thus far could be safely disposed of. It also urged that plans for safe disposal of radioactive wastes be made a prerequisite for approving the site of any future installation by the AEC or under its jurisdiction, and that the AEC consider concentrating its chemical processing activities at a minimum number of sites located in satisfactory places. The committee which authored this letter had a distinguished membership. It was chaired by the late Harry H. Hess, chairman of the geology department at Princeton University, who was a member of the NAS, and it included two other NAS members as well, namely Hubbert and Richard J. Russell.*

The AEC delayed six months before answering, then gave the committee a polite brush-off, commenting that the committee's proposals were costly and unnecessary.[7] In 1970, the agency elaborated on this theme still further. "To comply with the Committee's recommendations," it said,

> AEC would have had to abandon fuel reprocessing and radioactive waste management facilities and activities at each of the above sites [Hanford, Savannah River, and NRTS in Idaho]. It would have had to acquire an extensive new site or sites, presumably located over either salt beds or deep synclinal basins, since such locations appeared most attractive to the Committee for disposal of waste. It would have had to construct new fuel processing facilities and waste management facilities at the new site and move existing radioactive wastes from existing sites to the new site for disposal. Such an undertaking would have involved the expenditure of billions of dollars.[8]

The AEC was apparently so miffed that it stopped using the committee. Thereafter, according to Hubbert, the committee had "practically no further duties except for trivialities." Until 1963, that is, when Hubbert became chairman of the earth sciences division of the NRC and promptly set in motion events that were to bring the committee into direct confrontation with the AEC. "I told them I didn't propose to keep any committee standing around twiddling its thumbs," he recalls. "I said they should either discharge it or give it something worthwhile to do." The AEC, with some misgivings, agreed that the committee should undertake a review of the waste disposal research and development program of the Division of Reactor Development and Technology, the unit of the AEC which the committee had been advising. The reactor division is concerned with radioactive wastes resulting from the nuclear power industry; other parts of the agency are responsible for the wastes generated by the AEC itself.

_____

*The other members included John N. Adkins, William E. Benson, John C. Frye, William B. Heroy, and Charles V. Theis. William Thurston served as secretary. Three members of the committee—Russell, Hess, and Adkins—had served as chairmen of the NRC earth sciences division, while a fourth (Hubbert) would later serve in that post.

In April and May of 1965 the committee visited the principal AEC disposal sites—Hanford, Savannah River, NRTS in Idaho, Oak Ridge, and experimental salt mine facilities in Kansas.* At each site the committee was briefed on research sponsored by the reactor division. It also inspected the actual operating facilities used by contractors to dispose of the AEC's own wastes. Afterward, the committee drafted a report that discussed not only the research program of the reactor division, but also the operating procedures of AEC contractors—a task it had not been assigned. The committee explained that it ventured beyond its "specific delegated responsibilities" to the reactor division and concerned itself with "all phases of ground disposal of radioactive wastes" because, "like all responsible citizens, the members of the Committee are concerned for the welfare of man and the perpetuation of an environment in which he can satisfy his physical needs and realize his cultural aspirations."[9]

The report was highly critical of the waste disposal practices at the major AEC installations. As Earl Cook, the group's executive secretary, later summarized the committee's conclusions: "It had become clear that (1) not one of those sites was chosen with safe waste disposal in mind; (2) not one of the sites has proved capability for safe disposal on site of all waste produced by the plant; and (3) compromises between safety (defined as isolation from the biosphere of all radioactive wastes during their hazard lives) and economic expediency had been and were still being made."[10]

When the committee sent a draft of its report to the sponsoring agency (a frequent practice among Academy commit-

---

*The committee was now chaired by John E. Galley, a consulting petroleum geologist from Texas. The other members were Charles W. Brown, a research geologist with Socony Mobil Oil Co.; George B. Maxey, research professor in hydrology and geology at the University of Nevada's Desert Research Institute; John C. Maxwell, chairman of the geology department at Princeton University; Charles Meyer, professor of geology at the University of California at Berkeley; Robert C. Scott, a water resources specialist with the U.S. Geological Survey; Charles V. Theis, research hydrologist with the U.S. Geological Survey; and A. F. Van Everdingen, a petroleum engineer with DeGolyer and MacNaughton, a Dallas, Texas, consulting firm. J. Hoover Mackin, chairman of the NRC Division of Earth Sciences in 1965, a geology professor at the University of Texas at Austin, was an ex officio member of the committee. Earl Cook, former director of the Idaho State Bureau of Mines, served as executive secretary.

tees), the AEC immediately tried to get the committee to delete its criticisms of the operating practices of AEC contractors.[11] But the committee unanimously refused to delete anything—it agreed only to recast its criticisms as background to recommendations for further research. The report was finally completed in May 1966; its text contained the harshest criticisms yet leveled at the AEC's waste management program.

The report reiterated the committee's continuing conviction that "none of the major sites at which radioactive wastes are being stored or disposed of is geologically suited for safe disposal of any manner of radioactive wastes other than very dilute, very low-level liquids."*

With respect to all of the ground-disposal procedures then in routine operation, the committee acknowledged that "no serious hazards" had been created "at present." But it expressed concern "about the long-term safety of the operations if they are to be continued at the same sites for many decades or even for centuries." The committee also faulted "the working philosophy of some operators, although certainly not that of the AEC, that safety and economy are factors of equal weight in radioactive-waste disposal." It noted that "lack of funds has been cited at one location or

---

*The only exception cited by the committee was a hydrofracture-and-grouting technique used to dispose of intermediate-level liquid wastes at Oak Ridge. The report found fault with waste disposal practices and plans at virtually every site. With respect to NRTS in Idaho, the committee expressed anxiety "that considerations of long-range safety are in some instances subordinated to regard for economy of operation" and that "some disposal practices are conditioned on over-confidence in the capacity of the local environment to contain vast quantities of radionuclides for indefinite periods without danger to the biosphere." With respect to Hanford, the committee expressed concern "over the prevailing belief" that the top layers of several hundred feet of soil, sand, and gravel would "provide a reservoir for safe storage of tremendous quantities of wastes of all levels of radioactivity, and that no hazardous amounts of radioactivity will percolate down to the water table." With respect to Savannah River, a majority of the committee argued that a plan to bury high-level wastes in bedrock deep beneath the site was "in its essence dangerous." And with respect to Oak Ridge, the committee reiterated its opposition to the use of seepage ponds. The committee was generally opposed to the existing practice of putting intermediate- and low-level liquid wastes as well as solid waste directly into the ground above fresh water zones. It acknowledged that such practices were "momentarily safe" but warned that they would "lead in the long run to a serious fouling of man's environment."

another for inability to conduct needed research or to use alternate disposal methods which are agreed to be safer than current practices." But it added: "It is apropos to point out that waste-disposal costs are now a small part of the overall expense budget of the nuclear industries, and that any compromise with safety for the sake of economy could lead, in the long run, to a mushrooming of waste disposal into the most costly item in the use of nuclear power."[12]

The report—the toughest attack on the waste disposal program yet issued—was transmitted to the AEC in May 1966; it promptly disappeared from sight for several years. On November 7, 1966, the AEC sent a fifteen-page critique to the Academy which purported to show that the committee's report had been misguided in its major conclusions and recommendations and inaccurate in various details. In a cover letter, the AEC said that, since the report had already been made available to pertinent personnel, "we do not believe that additional distribution or publication of the report is warranted."[13] So the report was suppressed. Its disturbing conclusions were made known only to a handful of insiders despite repeated efforts by the committee to get the document released. The committee prepared a rebuttal of the AEC's critique but this, too, was never made public.*

What's more, the committee itself was disbanded. The

---

*The committee was not the only part of the Academy to express concern over AEC waste disposal practices. Even before the committee had completed its report, the Academy convened a separate ad hoc group headed by Academician Abel Wolman, professor of sanitary engineering at the Johns Hopkins University, which concluded:

Because the AEC is an operating agency, there has been a tendency to solve storage and disposal problems on an *ad hoc* basis. There is a need for a long-range, comprehensive plan that will elucidate the principles and practices needed to solve not only present problems but those of the future; the plan should take into account the possible effects of unusual natural events and disasters, as well as foreseeable man-related environmental changes; and it should reflect an awareness that expedient small-scale practices may be hazardous, particularly with respect to long-lived nuclides, if the practices continued to be carried on for a long period of time or on the enlarged scale expected to be reached in 1975. [Frederick Seitz to Glenn T. Seaborg, August 30, 1965.]

But the AEC brushed aside the Wolman group's concerns with a reply that contended the AEC already had matters well in hand (Glenn T. Seaborg to Frederick Seitz, November 1, 1965).

AEC, which had been funding the committee's work under contract, informed the Academy that it intended to end the advisory relationship as of mid-1967 "because a major part of our ground disposal R&D program is reaching a successful conclusion."[14] That meant that the committee could no longer function unless it found another source of support. The Academy made no effort to obtain additional support because, as then-President Seitz recalls, "It isn't easy to get support for a committee so obviously related to the work of any agency.... You'd have had to beat a lot of bushes."

The possibility that the AEC might cut itself off entirely from any independent advice alarmed the top leadership of the Academy. "Everything I have seen during recent years suggests to me that the geologic problems associated with radioactive waste disposal will grow rather than diminish in the period ahead," Seitz told the AEC. " ... As a result I would have substantial hesitation about terminating the committee without assurances that the gap left would be appropriately filled."[15] Further meetings were held between Academy and AEC officials and a compromise was finally reached, though the price to the Academy was high. The original committee was dismissed, and a new committee was established with a virtual guarantee that the AEC would have closer control over its operations. A proposal submitted to the AEC by the Academy on February 29, 1968 (and accepted by the AEC on March 11, 1968), stated that membership of the new committee "shall be discussed with the AEC"—thus giving the agency an implied veto over the makeup of the committee. The proposal also said that committee reports would be furnished to the agency "for its consideration and any distribution beyond the AEC"—thus acknowledging that the AEC would retain the right to suppress reports that were not to its liking.[16] Finally, the new committee was placed in the NRC Division of Chemistry and Chemical Technology, a move which had the effect, intended or not, of putting it under the supervision of a division which had closer ties to the nuclear industry than did the earth sciences division under which the old committee had operated. The chairman-designate of the chemistry division at the time the committee was formed was an executive of the DuPont Company, which operates the AEC's Savannah River Plant; one of the members of the division's executive committee was a vice-president of the General Electric Company, which had operated the AEC's Hanford plant from 1946 to 1966 and which was building a nuclear fuel reprocessing plant that would generate high-level wastes.

Thus the AEC had gained a more controlled advisory relationship under the new set-up. The only real gain to the Academy was that the new committee—called the Committee on Radioactive Waste Management—would have a broader mandate than the old. It would advise the agency as a whole, not just the reactor division. And it would deal with the broad issue of waste disposal, not just the geologic aspects of the problem. Moreover, it would be concerned with operational matters, in addition to the research aspects which were supposed to be the main focus of the previous committee.

Nevertheless, the circumstances suggest that the AEC emerged with the upper hand. The new committee was loaded with scientists who had close ties to the AEC or its major contractors. The committee's first chairman was Clark Goodman, head of the physics department at the University of Houston, who had formerly served as assistant director of the AEC's reactor division, the very division which had been at war with the previous committee. The new committee also included a former deputy director of that same division; a former manager of the AEC's Hanford Laboratories; a former atomic energy official of the DuPont Company, which operates the AEC's Savannah River plant; and a former deputy director of the AEC's Brookhaven National Laboratory. The four remaining members, although based at universities or state agencies, had ties with the AEC ranging from long-term consulting arrangements to a brief period spent as a visiting scientist at an AEC installation.*

---

*The members of the committee, in addition to Goodman, included: Robley D. Evans, Massachusetts Institute of Technology, a longtime consultant to the AEC; John C. Frye, chief of the Illinois State Geological Survey, a member of the health physics advisory committee of the AEC's Oak Ridge National Laboratory; Jack E. McKee, California Institute of Technology, a member of the AEC's Advisory Committee on Reactor Safeguards; Herbert M. Parker, Battelle Memorial Institute, former manager of the AEC's Hanford Laboratories; Louis H. Roddis, Jr., General Public Utilities Corporation, former deputy director of the AEC's reactor development division; John H. Rust, University of Chicago, who had spent a year as a visiting scientist at Oak Ridge; Clarke Williams, former deputy director of the

In contrast to the cozy relationships between the new committee and the AEC, the previous committee had had very few members who were close to the agency, partly because it was composed of geologists and the AEC has never had much expertise in geology, partly because it strove to remain independent. In fact, Hubbert, when he became chairman of the NRC earth sciences division in 1963, made a point of dropping one member of the old committee, the late William B. Heroy, after it became known that a consulting firm with which he was associated had a contract to advise the AEC on waste disposal in salt.

Although the AEC seemingly had a profound influence over the committee in its formative years, that influence has since slackened. Academy officials insist that the AEC no longer has the right to suppress Academy reports; they say the Academy itself now decides whether a report will be made public.

The complexion of the committee has also changed over the years so that now it is less blatantly loaded with former AEC officials. But an AEC flavor remains. Of the twelve members of the committee during 1972–73, five had once held key positions with the AEC, its laboratories, or its industrial contractors,* while some of the others are said to

---

AEC's Brookhaven National Laboratory; and Hood Worthington, consulting nuclear engineer, who had served as an atomic energy executive for the DuPont Company, which operates the AEC's Savannah River Plant, and who had served on AEC advisory panels. The staff man for the committee was Cyrus Klingsberg, who had worked for the Office of Naval Research before coming to the Academy. The AEC's liaison man was initially John A. Erlewine, who was then assistant general manager of the agency.

These ties may be but the tip of the iceberg. Since an agency's grants, contracts, and consulting arrangements are seldom publicized widely, it is difficult to ascertain what relationships exist between a given scientist and a given agency.

---

*The five are W. Kenneth Davis, vice-president of the Bechtel Corp., former director of the AEC's reactor division and former president of the Atomic Industrial forum; Herbert M. Parker, consultant and former manager of the AEC's Hanford Laboratories; F. H. Spedding, former director of the AEC's Ames Laboratory at Iowa State University; Clarke Williams, former deputy director of the AEC's Brookhaven National Laboratory; and Warren F. Witzig, head of the nuclear engineering department at Pennsylvania State University, who has held positions in the nuclear industry and at the AEC's Bettis Laboratory.

have received AEC financial support for their research projects.

The new committee has issued three reports since it was formed in 1968. Each has been supportive of the AEC although none can be described as a complete whitewash.

The first, produced in February 1970, was labeled an "interim report." It is a skimpy nine-page document which is of interest primarily because it served the AEC's political purposes so well. Late in 1969 the AEC came under pressure to release the Academy's 1966 report that had been so critical of AEC waste disposal practices. On October 7, 1969, Senator Frank Church (D-Idaho), whose home state includes the AEC's National Reactor Testing Station, informed the agency he knew of the existence of the report and asked why it had not been made public. In reply AEC Chairman Glenn T. Seaborg implied the previous committee had made numerous errors and had therefore been replaced with a more competent committee by the Academy.[17] (He neglected to mention that the previous committee was dissolved only because the AEC had cut off its funding.) This new "broader" committee would prepare a report of its own, Seaborg promised. But Church was not satisfied. And in March 1970, Senator Edmund Muskie (D-Maine), whose subcommittee on air and water pollution was holding hearings on underground uses of nuclear energy, formally asked the AEC for a copy of the 1966 report.[18] At that point, the AEC gave in. The agency made the report available to the press and explained that it had not really "suppressed" the report; it had simply not published it (conveniently forgetting that it had also not allowed the Academy to publish it).[19] Once again, the AEC promised that the new committee, with its "broader spectrum of scientific disciplines," would be reporting its impressions soon.[20]

It is not surprising that the new committee's "interim report" gave the AEC what it needed. The report was little more than a long list of everything the committee had done in its first two years—meetings held, AEC sites visited, and topics discussed. But it contained one value judgment which seemed to run counter to the findings of the previous Academy committee. "The Committee noted the extensiveness and care in waste management at each site visited," the interim report said. "The Committee is gratified by the quality and scope of the R&D program sponsored by the AEC in radioactive waste management."[21] The AEC submitted the interim report to the Muskie subcommittee along with the suppressed 1966 report. Thus the AEC had managed to use the new committee to blunt the criticisms raised by the old.

The new committee's second report, entitled *Disposal of Solid Radioactive Wastes in Bedded Salt Deposits*, proved to be a minor embarrassment to the Academy. It was published in November 1970 after a hectic last-minute flurry of rewriting. The committee had originally been asked to review the general concept of burying wastes in bedded salt. As noted above, the previous Academy committee had first suggested burial in salt, though it had not specified what types of salt formation should be used. Subsequent studies conducted by Oak Ridge National Laboratory, with periodic advice and encouragement from the previous Academy committee, had indicated that bedded salt (long horizontal layers) would be preferable to salt domes (narrow vertical formations) and that the liquid wastes could be solidified before burial, thus reducing the possibility that the wastes would migrate from the burial spot. After the previous Academy committee had been dismissed, Oak Ridge had further developed the technology and plans for salt burial, so the new committee was asked to review the concept once again before the AEC progressed further with the project. A panel of experts appointed by the committee visited Oak Ridge in May 1970 and held a meeting in Lawrence, Kansas, in June (Kansas has extensive salt deposits and had been the site of several Oak Ridge experiments). The panel was actually working on the second draft of a report reaffirming the salt concept and was in the midst of its meeting in Lawrence when the AEC pulled a surprise. John A. Erlewine, assistant general manager of the AEC, held a press conference in Topeka and announced the "tentative selection" of a salt mine near Lyons, Kansas, as the site for a project to demonstrate the feasibility of burying radioactive wastes.[22] The mine had long been discussed as a possible national repository for wastes generated by the commercial nuclear industry and, if necessary, for AEC-generated wastes as well. Erlewine was the AEC's officially designated liaison to the Academy committee, but he had not told the committee of the agency's latest plans. The announcement caused consternation on the panel, which felt it could not very well come out months later with a report that discussed salt but made no mention of the AEC's specific plans. So the panel tried to shift its focus late in the game. As Frye, who was a member of the panel as well as chairman of the parent committee, explained in an interview: "We started out considering the general concept of bedded salt and then, kind of at the last minute, tried to focus it on one site because the AEC said 'That's where we're gonna go.'"

The panel revised its report, and the parent committee then rewrote the revision substantially. As finally published, the report had two major conclusions:

1. The use of bedded salt is "satisfactory" and, in fact, is "the safest choice now available, provided the wastes are in an appropriate form and the salt beds meet the necessary design and geological criteria."
2. "The site near Lyons, Kansas, selected by the AEC is satisfactory, subject to the development of certain additional confirmatory data and evaluation."[23]

The report was not an all-out endorsement of the Lyons site. It suggested a number of studies that should be completed before any radioactive materials were actually committed to the salt beds and, like many Academy reports, it included numerous caveats—in this case, to cover the committee's flanks should the site prove a bad choice. But the tone of the report suggested that the Lyons mine was suitable and that the additional studies would not constitute a serious impediment to proceeding with actual burial. As the report expressed it, "Based on research and development performed to date the Committee does not anticipate any insurmountable problem."[24]

Proponents of the Lyons site welcomed the Academy report as an endorsement of their views. Milton Shaw, director of the AEC's Division of Reactor Development and Technology, asserted in Congressional testimony that the site "had been recommended" by the Academy.[25] And Chet Holifield (D-California), a strong supporter of the AEC, triumphantly told a critic of the site: "The Committee on Radioactive Waste Management of the National Academy oₓ

Sciences said to go ahead. You heard here the two recommendations they made. You heard them the same as I did."[26]

The Academy report would apparently have been even more enthusiastic about the Lyons site had it not been for the influence of a single member of the panel, William W. Hambleton, director of the Kansas Geological Survey. Hambleton had been appointed to the panel after he had recommended to the governor of Kansas that site selection for the waste repository be deferred pending further studies to demonstrate the integrity of the Lyons mine. Hambleton and the Survey became the focal point of scientific opposition to the AEC's plans. They did not flatly reject the Lyons site, but argued that the safety of salt in general and of the Lyons site in particular had not been adequately established. Hambleton issued a series of reports in Kansas criticizing the AEC for its failure to conduct various studies he felt were needed. Meanwhile, he lobbied intensively to persuade the Academy panel to raise questions about Lyons.

As it turned out, the Lyons site was not suitable. In September 1971 the AEC revealed that two unexpected discoveries cast doubt on the feasibility of using the Lyons mine. One discovery, which resulted from a survey conducted by a consultant hired by Oak Ridge, was that there are at least twenty-nine abandoned gas and oil drillholes which extend into or below the salt formation near the site. The consultant concluded that, while twenty-six of these could probably be plugged successfully, the likelihood of plugging the other three was "very low."[27] Thus the possibility was raised that water could leak from these abandoned holes into the Lyons mine, destroying its integrity. The second discovery was that an adjacent salt mine, operated by the American Salt Company, had made extensive use of "solution mining," a technique in which water is used to dissolve the salt. Such mining leaves no supporting pillars underground, thus introducing what the AEC called a "potential for sudden and dramatic collapse of a fairly large area not too far from the Repository site, with the formation of a surface lake which could be several hundred feet deep." The AEC claimed such a lake would probably have "no real technical significance to Repository safety" but said its formation "could certainly engender unfavorable emotional and public relations problems." The situation was complicated by the fact that American Salt had once used a hydraulic fracturing technique in which water is forced down one hole where it cracks the salt bed and then works its way over to a second hole and returns to the surface carrying dissolved salt. American Salt tried the technique once in the mid-1960s. It worked well for a brief period, but then some 175,000 gallons of water suddenly disappeared. Neither the company nor the AEC knows where it went. Thus the possibility exists that the missing water, and other water from the solution mining as well, could be migrating toward the proposed waste disposal site. To make matters worse, American Salt revealed that it planned to double its solution mining activities.

The sudden revelations forced the AEC to announce that, pending further study, "We are holding in abeyance any further site oriented work at Lyons, including leasing of land and plugging of holes."[28] The AEC promptly asked the Kansas Geological Survey and the Oak Ridge drilling consultant to search available records for other possible sites. The results of that survey were made public on January 21, 1972. Of eight areas considered, three were judged to have potential worth investigation, four were deemed less promising, and the area which included Lyons was deemed "the poorest candidate" of all, largely because it has numerous old oil and gas holes, a large number of producing wells, water above and below the salt, possible deep-seated structural problems, inadequate buffer zones, and high potential for development of oil and gas reserves in the future.[29] The setback at Lyons forced the AEC to announce, in May 1972, that it planned to build engineered surface facilities to store high-level wastes produced by the nuclear power industry.[30] Such facilities would give the AEC an alternative should no suitable geologic repository be found. Meanwhile, the AEC said it would also search for suitable locations in salt and other geological formations outside Kansas and would explore longer-range concepts such as disposal in space, disposal under the polar ice caps or the sea bed, and conversion of the toxic material by nuclear processes called transmutation.[31]

The Academy report did have the foresight to recommend that a survey be made of oil and gas wells in the surrounding area, but made no mention whatever of solution mining.

Why didn't the committee itself envisage the problems at Lyons? A number of factors probably conspired to lull the Academy into endorsing the AEC's plans. The last-minute change of focus, in which the panel suddenly decided to discuss Lyons as an afterthought, virtually guaranteed that the discussion would be superficial rather than a full-scale site analysis. The AEC orientation of some members of the parent committee, which rewrote the panel's report substantially, may have influenced the report's optimistic tone. The Academy's long-standing support for the use of salt beds may have made the panel a bit too eager to get on with an actual demonstration project. And the part-time nature of Academy committee work made the Academy largely dependent on the information supplied by others. If the committee had known about the numerous drill holes and the solution mining, it would presumably have been more skeptical of the Lyons site. But it was not apt to uncover such information independently. "Anybody's got rocks in their head if they expect guys working for no salary to do the kind of job Bob Walters [the Oak Ridge consultant who found all the drill holes] did on a damn good retainer," says John Frye, Academy panel chairman.

The difficulties that arose at Lyons led Senator Mike Gravel (D-Alaska), a frequent critic of the AEC, to complain that "The failure of both the AEC's final environmental [impact] statement and the National Academy of Sciences review to disclose such serious problems at Lyons, Kansas, raises a mammoth question. What good is [an impact] statement, or a NAS review?"[32] Academy officials consider that phraseology overly harsh. "Gravel's a horse's ass," says Klingsberg. "I think the Academy and this committee covered themselves professionally with great distinction." But the circumstances surrounding this particular report suggest that Gravel may be essentially correct: The committee seemed just a bit too willing to endorse an AEC plan which had not been adequately researched.

The committee's third report, published in early 1972, was an evaluation of a plan to store high-level wastes in caverns mined in the bedrock beneath the Savannah River Plant in South Carolina. The plan had first been proposed to the AEC in 1959 by the DuPont Company, which operates Savannah River under contract; it had been under continuous investigation ever since. The high-level wastes generated at Savannah River thus far (largely as a byproduct of weapons work) have been stored in underground tanks, but this is considered only an interim solution until a more permanent disposal site can be found.

The previous committee had been cool to the bedrock concept from the start. Hubbert recalls that the previous committee objected that the bedrock would inevitably be criss-crossed by joints and fractures which would allow the wastes to leak out, thus endangering the Tuscaloosa aquifer, a major source of drinking water for much of South Carolina and adjacent parts of Georgia. The water-bearing Tuscaloosa Formation lies directly beneath the Savannah River Plant and directly above the bedrock into which the wastes would be deposited. Another concern was that the wastes might migrate to the nearby Savannah River.

The committee had considered the bedrock proposal in some detail in its 1966 report, which was suppressed. At that time a majority of the committee concluded that the plan was so inherently hazardous that all further work on the project should be stopped. "The placement of high-level wastes 500 to 1,000 feet below a very prolific and much-used aquifer is in its essence dangerous and will certainly lead to public controversy," the majority said. "Any demonstration of safety must leave no shadow of doubt." The majority concluded that "apparently, the only safe disposal for high-level wastes would be an offsite disposal, presumably involving solidification before transportation." However, a minority of the committee recommended that work continue and that various steps be taken to further test the concept.[33]

The AEC chose to ignore the majority opinion and instead followed the minority recommendations. As the agency later explained, "Even with an assumed poor probability of carrying the bedrock program to completion, the potential savings in public funds was estimated to be so great that continuing the program to a definite end, one way or the other, seemed warranted." The AEC estimated that it would cost $500 million to solidify the wastes and ship them to an offsite repository for permanent storage, compared with only $80 million to put them in the bedrock. Meanwhile, the DuPont Company, dissatisfied with the Academy's report, had appointed its own panel of scientific consultants. They reported in May 1969 that the bedrock concept was sufficiently promising to warrant construction of an access shaft and several exploratory tunnels. Such tunnels were essential to disclose the extent of fissures and fractures in the bedrock, the DuPont panel said. "The probabilities of producing evidence to warrant the completion of the entire project are high."[34]

That was where matters stood when the new Academy committee was asked in early 1971 to review all previous reports and all newly acquired data. The committee appointed a panel which interviewed project officials as well as critics from the previous Academy committee. In early 1972 the new committee published a report which concluded that there is "a reasonable prospect" that wastes can be stored in the bedrock and kept isolated from the biosphere for at least a thousand years. The panel concluded that no reasonable amount of exploration from the surface could conclusively demonstrate the safety of the concept, so it recommended that an exploratory shaft and tunnels be sunk. But the report warned that, if data from these explorations did not "clearly confirm" the safety of the project, then the concept "would become invalid."[35]

The new Academy committee had thus disagreed with the majority on the old committee, which had recommended dropping all further investigation of the bedrock concept. The disagreement stemmed partly from the use of different assumptions and calculations to determine how fast the waste is apt to migrate from the burial spot, partly from the fact that the new committee was looking at more recent information, and partly from differing assessments of the AEC's intentions. The old committee feared that the AEC would make a decision based primarily on economics without due regard for public safety. It did not really oppose exploratory tunnels per se but rather feared that such tunnels would simply increase the momentum behind a project it regarded as inherently dangerous.

The new committee, on the other hand, is less suspicious of the AEC. Klingsberg is satisfied that "the social conscience of the AEC is first-rate." And he believes that, if the AEC tried to go ahead with something dangerous, the Academy could always block it. "If the Academy committee can build up pressure to go ahead," he says, "then the Academy

committee can build up a hell of a lot of pressure to stop it."
Maybe so, but the performance of the committee to date
suggests that it would tend to go along with the AEC in the
absence of overwhelming evidence to the contrary.

However, on November 17, 1972, the AEC itself had second
thoughts. Faced with budgetary problems, political opposition in South Carolina, and the realization that it would
require "considerable additional investigation" to demonstrate the safety of disposal in the bedrock, the AEC
announced that it would "indefinitely postpone further
development of the bedrock project" and would place priority on "other disposal methods."[36]

The story of the radioactive waste committees highlights
a dilemma that continually confronts the Academy: How
can a committee maintain a long-term advisory relationship
with an agency, a relationship which depends on mutual
trust and agency funding, and yet remain independent,
even critical, in viewpoint? The old committee tried to adopt
the role of independent gadfly. It successfully pushed the
salt concept, but then it became, in AEC eyes, a bothersome
nuisance when it repeatedly charged that none of the AEC
plant sites was suitable for waste disposal. The AEC disagreed, and it exercised its prerogative of ignoring the
advice. The committee waged its campaign vigorously
behind the scenes, but its nagging did little more than anger
the AEC—for, while the committee itself was willing to
challenge the AEC publicly, the Academy leadership at the
time was not. When conflict arose over the 1966 report,
President Seitz chose not to make an issue over the disagreement. Instead, he bargained for a compromise that
would allow the Academy to continue operating as a close
adviser within the AEC orbit. Unfortunately the new committee which emerged from that compromise abandoned the
gadfly role. It drew many of its members from among scientists who are close to the AEC. And it acted more like a
consultant serving a client than like an independent critic.
The new committee is by no means unwilling to raise questions about AEC projects, but it does so as a family adviser,
not as a disinterested examiner.

This approach has its advantages. Klingsberg says the
AEC is now more willing to listen to what the Academy says

and that the exchange of information between the two organizations is smoother. Moreover, the AEC says it has benefited greatly from the committee's advice.

But the approach also has its drawbacks. For one thing, advisers who get too close to an agency find it difficult to adopt the probing, aggressive stance that is often needed to expose weaknesses in an agency's plans. Almost all the expertise on radioactive waste disposal lies in the hands of the AEC or its contractors, a circumstance which makes it particularly important to have a truly independent scientific assessment, even an adversary assessment, of AEC programs. In the case of Lyons, the Kansas Geological Survey adopted an adversary role. But in the case of Savannah River there has been no organized outside scientific assessment, either independent or adversary. The Academy, in theory, is supposed to supply an independent judgment. But the new committee's performance thus far—its production of an interim report which helped the AEC blunt the criticisms of the suppressed 1966 report, its willing endorsement of the Lyons site, and its support of bedrock exploration which the previous committee had condemned—all these suggest that the Academy is acting more as an accessory to AEC plans than as a hard-nosed guardian of the public interest.

The committee has also found it difficult to exert influence on matters the agency doesn't want discussed. It is generally agreed, for example, that the AEC has not put enough effort into solving the waste disposal problem. Two reports by the General Accounting Office in 1968 and 1971 attest to this.[37] The old committee tried to goad the AEC into devoting greater resources to the problem. Indeed, the old committee ran into trouble with the AEC because it exceeded its mandate and tried to comment on the AEC's whole approach to waste disposal. But the new committee does not consider this its job. "You really can't tell them what they should do except in a specific context," says Frye, "because if you say, 'Look—you're not putting enough effort into waste management in general,' they'll say 'Who the hell asked you?' and with some justification."

That statement underlines the subservience of the Academy's position and the timidity of its attitude. The new committee has done little more than comment on the feasibility of plans for specific projects—projects that were generally well developed before the committee's advice was even sought. It has never attempted a comprehensive review of the radioactive waste problem or of the AEC's program to cope with it. Nor has it compared the risks and benefits of burying wastes at one proposed site with the risks and benefits of disposing of the waste by other means. The AEC is supported by public funds. If it needs more funds for "safe" disposal, or if waste hazards threaten the expansion of nuclear power, then the public should be informed of the risks and Congress should debate the problem extensively. Unfortunately, the Academy, *the only technical group outside the AEC which has maintained continuing surveillance over the program*, has done little to raise the pertinent issues. One suspects that the AEC would be further along toward solution of the waste disposal problem if the Academy had found some way to alert the public to the magnitude of the problem and the inadequacy of the attention being paid to it.

The Academy has made genuine efforts in recent years to increase the committee's independence. It has added members who have no ties to the AEC. And, in early 1973, it placed the committee under the jurisdiction of the Environmental Studies Board, a move which should increase the committee's sensitivity to the environmental hazards of waste disposal. But these steps are merely palliatives. A key lesson to emerge from our examination of this committee is that the Academy's part-time advisory apparatus is not strong enough to keep effective watch over an agency whose operations involve incalculable risks to future populations. The nation needs a full-time, government-sponsored monitoring group of highly qualified scientists, totally independent of AEC funding and influence. Part-timers, unpaid, relying on the AEC for data, dependent on the AEC for funds, just cannot do the job.*

---

*In early 1975, the AEC's nuclear development activities, including its waste management programs, were to be merged into a new Energy Research and Development Administration, responsible for all forms of energy, not just nuclear. Such an administrative reshuffling would not obviate the need for independent monitoring, for the AEC was expected to form the core of the new agency, its promotional instincts intact.

CHAPTER SIX

The SST:

Playing the Government's Game

"The Academy was being asked to do a 'white-wash job' on a publicly unpalatable undertaking."

> —M. King Hubbert, former member of the Academy's Governing Board, in an October 10, 1968, letter describing the Academy's sonic boom reports.

"The SST work was not the proudest accomplishment of the Academy but it was not disgraceful."

> —Harvey Brooks, former chairman of the Academy's Committee on Science and Public Policy, in an interview, July 20, 1971.

The controversial issue of whether this country should build a supersonic transport (SST), a commercial plane capable of flying faster than the speed of sound, caused the Academy much internal grief and dissension. The Academy never really dealt with the broad questions of whether such a plane should be built and, if so, whether it should be allowed to fly over land or only over the water. But the Academy did consider certain aspects of the SST's possible impact on the environment, notably the possibility that SST exhausts might cause catastrophic changes in the earth's atmosphere and that sonic booms might harm buildings or exposed populations. Its handling of one of those issues—the likely impact of the sonic boom—sparked an internal battle that rent the usually placid halls of the Academy for the better part of a year. Before the fight was over, an unprecedented demand had been issued by 189 Academy members that one of the Academy's own reports be repudiated. Details of this fight have been hushed up by the Academy leadership but much of the internal correspondence and several documents that were once considered "privileged" have been made available to us. As a result, the SST controversy provides an unusually well-documented opportunity to assess just how an Academy committee can fall captive to a government agency and thus end up serving that agency's interest as opposed to the broader public interest.

The Academy seems to have played no role whatever in the original governmental decision to build a commercial SST. The first production planes capable of routine supersonic flight had all been military, but as early as 1956 the National Advisory Committee for Aeronautics, predecessor of the current National Aeronautics and Space Administration (NASA), had launched a research program directed toward a supersonic commercial transport, and by 1959–1960 three agencies—NASA, the Department of Defense, and the Federal Aviation Agency (now the Federal Aviation Administration)—were proposing a national program to develop such a vehicle for commercial use. Proponents of a commercial SST viewed it as an important American entry in an international competition with the British and French, who had both announced plans to develop such a plane. The proposal for an American SST made little headway in the closing days of the Eisenhower administration, but it picked up momentum after President John F. Kennedy took office in 1961 and issued a call for suggestions on how to "get America moving again." In response, the FAA came back with still another recommendation for an SST program; the idea was endorsed by various high-level groups, including a Cabinet committee; and on June 5, 1963, President Kennedy announced the SST as a national objective. Prototypes of an SST were to be developed through a partnership arrangement in which government would provide most of the financing and industry would perform most of the work.[1] The Academy was not asked, nor did it venture an independent opinion, as to whether the project was desirable or not.

It was not until May 1964, in fact, that the Academy was asked to comment on any aspect of the SST program, and then the request came under circumstances which suggested that the Academy was brought into play as much for political purposes as for its scientific expertise. The point at issue involved the sonic boom, a phenomenon that had been recognized from the start as one of the most significant environmental hazards apt to result from extensive SST flights. The boom is a shock wave created by an aircraft moving at or above the speed of sound. The strength and character of the boom are chiefly determined by the shape and weight of the airplane, its distance above the ground, atmospheric conditions, and the type of maneuvering performed by the aircraft. Depending on circumstances, the boom can sound like a thunderclap, an artillery piece, or a firecracker—or it can be virtually inaudible.

Perhaps the most alarming aspect of the boom was its pervasiveness. Contrary to popular belief, the boom does not occur only when a plane passes through the so-called sound barrier. Instead, it is generally heard at all points along the flight path of a plane that is flying at supersonic speeds.

Thus a fleet of SST's criss-crossing the country would obviously subject vast areas and tens of millions of people to repeated boomings. What would be the impact of such booming? Would it damage buildings? trigger landslides? cause physical or psychological harm to humans? disrupt animal life? No one knew for certain, but research had been carried on for a number of years by NASA, the FAA, and the Defense Department when suddenly, in May 1964, the future of the sonic boom research program was thrown into doubt. At that time the FAA was directing, in Oklahoma City, Oklahoma, the most extensive sonic boom tests ever conducted. Air Force planes were subjecting the city to repeated booming and efforts were being made to determine the impact on buildings and on the population. Long before the five-month test series was completed, however, it had provoked a barrage of citizen complaints, lawsuits, claims for damage, and even a vote by the City Council (later rescinded) asking the FAA to suspend the experiment. The extent of the opposition was surprising in a city which housed the FAA's own headquarters installation; where one of every four jobs was in an aircraft factory, Air Force base, or other aviation-related activity; and where the city's commercial leadership hoped to establish the nation's first supersonic airport. But the FAA, with strong backing from the local chamber of commerce and the city's leading newspapers, had managed to ride out the storm—until May 12, 1964, that is, when a state court judge unexpectedly issued an order restraining various officials of the FAA and the Air Force from continuing the experiment. The court order (later quashed) caught the FAA by surprise. Not only did it jeopardize the testing program, but it came at a low point in the fortunes of the SST program. Just seventeen days earlier President Johnson had reluctantly announced that initial design proposals submitted by air-frame and engine companies had been judged inadequate, and the government was on the verge of making new design awards.

In a moment of panic, the agency turned to the Academy for help. The very day the court order was issued, an urgent telephone call was received by the Academy from the FAA Administrator's office. The caller wanted an advisory panel of respected experts assembled as soon as possible. Three days later the normally slow-moving Academy had a delegation, headed by Frederick Seitz, then president of the Academy, on hand at FAA headquarters for an initial briefing. Eight days later, on May 20, the Academy had a team of three engineers and two staff men in Oklahoma City.[2] The FAA took that occasion to announce that the Academy, "the nation's most distinguished scientific body," would conduct "an objective review of the sonic boom study now underway in Oklahoma City, Oklahoma." Beyond this initial review, the FAA said, the Academy would assemble a group of leading scientists and engineers to "analyze all aspects of boom phenomena, and provide advice and guidance for future research and study."[3]

The impact of announcing the Academy's participation, according to a perceptive account of the incident by John Lear, science editor of *Saturday Review*, was that it subtly shifted the onus of the SST project from the FAA to the Academy. As Lear saw it, the "bugaboo science" had been used to "frighten" the Oklahoma City Council, which had been a focal point for citizen opposition to the tests, "away from its proper exercise of political responsibility."[4] The essential correctness of Lear's analysis was revealed by the speed with which SST proponents seized upon the Academy's involvement as a means of countering critics of the program. Senator A. S. (Mike) Monroney (D-Okla.), a strong supporter of the SST who had personally suggested that the Academy be called upon for advice, immediately told prominent constituents who were clamoring for an end to the tests that the Academy was "well qualified to review all aspects of this experiment."[5] Similarly, Najeeb Halaby, then FAA Administrator, cited appointment of the Academy committee as an answer to demands by Oklahoma's other senator, the late J. Howard Edmondson, that the tests be suspended.[6] Halaby, in the opinion of the *Tulsa Tribune*, was hoping that "the prestige of the Academy will muffle some of the criticism coming from Oklahoma City."[7]

The use of a prestigious study group to head off political opposition is an old trick around Washington and is not necessarily reprehensible if the study group does a good job. But the Academy showed a disquieting tendency to place itself on the FAA's side of the controversy rather than act as an independent source of advice. M. King Hubbert, a distinguished geophysicist and member of the Academy, recalls that when the governing board of the National Research Council first considered whether the Academy should advise the government on the sonic boom problem, he tried to raise questions about the appropriateness of the Academy's role but "got the brushoff" from the Academy leadership and was denied details of the proposed contract. As Hubbert described it years later:[8]

"The whole affair was presented and handled in an atmosphere of secrecy and intrigue. So much was this the case that during the discussion I stated that it did not appear to me that the Academy's advice was being sought on what damage was likely to be produced by the booms from a supersonic transport, or whether such a transport should be built—that decision was apparently already a fait accompli—rather, the Academy was being asked to do a 'whitewash job' on a publicly unpalatable undertaking. All information on this subject which has come to me subsequently is consistent with that original judgment."

Despite Hubbert's fears, the project started off in a seemingly objective fashion. The initial Academy group of three engineers and two staff men visited Oklahoma City on May 20 and 21, appraised the test procedures used to measure the impact of sonic boom on structures, submitted a report recommending numerous improvements in those procedures,[9] then went out of business and was succeeded by a ten-man committee which would be responsible for carrying out a longer-range analysis of virtually all aspects of the boom problem.

The chairman of this new Committee on SST–Sonic Boom was John R. Dunning, dean of the school of engineering and applied science at Columbia University, who recalls that he was picked for the job because "I had no connection with the aircraft industry. . . . I had no ax to grind." The vice-chairman was R. G. Folsom, president of the Rensselaer Polytechnic Institute. The committee, both in its original form and as expanded in later years, seems to have encompassed a reasonable balance of professional expertise. It included members with backgrounds in aeronautical engineering who presumably would be sympathetic toward the development of new aviation technology. But it also included social scientists, psychologists, medical experts, housing specialists, a prominent conservationist, and even an insurance executive—all of whom would presumably be concerned with the well-being of people or structures and would have no visible reason to jump on the SST bandwagon. At least one member of the original committee—a former airlines official—publicly questioned the desirability of the SST shortly after he was appointed to the Academy group. And the top staff man in the early days of the committee, Richard Park, says he personally opposed the SST. Thus it is hard to argue, as some Academy members have, that the committee was "loaded" with SST sympathizers.

The committee's work, which was launched at a meeting on July 29, 1964, ultimately stretched over a seven-year period and consumed more than $500,000 in government support funds.[10] As is true of most Academy committees, the group did no original research on the sonic boom. Its function was to review the work done by others and to recommend further research. As the work progressed, the committee set up subcommittees to handle specific aspects of the sonic boom problem.

Although the appointment of committee members suggested that the Academy might be able to do an objective evaluation, the committee approached its task in such a way that it immediately became an ally of the FAA rather than an impartial source of advice. The committee, at least in its early days, handled its job for the government much as a lawyer handles a problem for his client. It sought to help the government promote its interests—in this case, the SST—and it did not worry much about whether those interests were desirable in some broader perspective.

This attitude became apparent when the Academy submitted its first two status reports to the government in January and July 1965. The wording of these two reports, which were not made public at the time, suggests that the committee regarded complaints over sonic boom effects as an annoying obstacle hampering the development of aviation rather than as a legitimate indication of a problem. The first status report, for example, laments that damage to poorly constructed buildings "may be attributed to booms, and thus—with the possible nuisance of making the repairs—may well continue, through the public, to *plague aviation.*"[11] (Emphasis added.) The second report warns that a boost in the number or severity of sonic booms "increases the *risk* [emphasis theirs] of an effective public protest, perhaps of a political character." The report explains that "The trigger for such a protest might well be some irrelevant accident or coincidence or the emergence of an aroused leader of hostile public opinion," but it fails to suggest that the trigger might also be a genuinely serious sonic boom accident.[12]

Both reports are written in the uneven manner that often characterizes committee efforts, with the result that it is possible to pluck out sentences here or there to support opposing points of view. But the giveaway of the committee's attitude comes in sections of both reports devoted to public attitudes. Incredibly enough, the committee seemed to feel that one of its responsibilities was to help the government plan a public relations strategy to persuade people to accept the SST program. In 1965, the committee even appointed a public relations consultant to its staff to help on this aspect of the work.[13]

The PR campaign, as envisioned by the Academy, was to be a soft-sell effort to "educate opinion leaders" so that they, in turn, would help develop "attitudes sympathetic to the problem" among the general public. "This would not be in any sense a 'campaign to sell the public on the SST,'" the committee insisted, "but rather a policy of presenting the facts to the public, of dispelling false ideas and unfounded fears, of urging the public to avoid premature conclusions based on fragmentary information rather than solid facts." Then the committee went on to recommend something that sounded suspiciously like just such a selling campaign. It implied that the protests in Oklahoma City had occurred in part because "there was not a great deal of public warning and preparation"; it therefore recommended that a "pilot public information program" be launched before any future tests in another city. "Such a program," the committee said, "would involve contacts with city leaders, editors and commentators to explain to them what is planned, to show them reasons why there should be no alarm or hysteria. . . ."[14]

This pilot program was to be followed by a long-range education effort under the leadership of "those agencies, organizations and commercial airline companies directly interested." The committee suggested that "The military agencies would wish to show the necessity for the supersonic plan for national defense"; airlines would wish to "show the advantages [of the SST] as outweighing objections"; air-frame manufacturers "would undoubtedly be eager to show engineering and design steps to increase the safety, efficiency, and total acceptability of the SST and to reduce sonic boom"; government agencies and airlines "could show what they were doing to increase operational effectiveness through route changes and flying patterns"; and "architects and contractors might subtly indicate what they were doing to sound-proof and strengthen windows, houses and other buildings."[14]

Presumably the FAA, which had been conducting a rather hard-sell campaign for years, didn't need such advice, but the mere fact that it was offered illustrates how far the Academy committee deviated from the role of impartial adviser. Instead of assessing the impact of the sonic boom from a neutral position, instead of limiting its attention to such questions as whether the boom would or would not be acceptable to the public, the Academy committee viewed its task as one of eliminating barriers to SST development, whether such barriers might be "irrelevant" public protests against the boom or a "plague" of damage claims. That a supposedly independent scientific body should fancy itself a public relations adviser for the government illustrates how far the Academy group had fallen.

For the first several years of its life, the committee operated in relative obscurity, submitting its status reports in private to the government (primarily to a high-level committee, chaired by the Secretary of Defense, which took over management of the program from the FAA) and making verbal contributions to government thinking through such devices as having its key members sit in with a coordinating group on sonic boom problems organized by the White House Office of Science and Technology. But starting in late 1967 the committee began issuing public reports and thus had a chance to contribute to the growing public debate over the controversial plane. Unfortunately, its performance created more confusion than enlightenment.

The first public report, which was completed in October 1967 but not issued until January 1968, dealt with the "generation and propagation of sonic boom." It offered scant hope that the boom could be eliminated—a position which tended to strengthen the hand of those who were opposing the SST while weakening the hand of the FAA. This particular report shows that one cannot always predict from the makeup of a committee how tough its final report will be, for, if ever a group should have been biased in favor of the SST, this was it. The report was prepared by a subcommittee on research, then approved by the full sonic boom committee. The subcommittee was chaired by Raymond L. Bisplinghoff, who had served in key positions related to the SST program in NASA, the FAA, and the Department of Transportation and who later received the FAA's Extraordinary Services Award for his contributions to the SST program. The other members of the subcommittee included two aeronautical engineers associated with universities, plus technical experts from Boeing, the prime SST contractor, and NASA, while the staff man was a retired Air Force colonel. All six had dedicated their lives to furthering aircraft development and thus presumably would look favorably on the development of an SST. Yet, when faced with the question of whether the sonic boom could somehow be "designed out" of the SST, the subcommittee came back with a pessimistic answer: "At the present time, future prospects for dramatic reductions in the intensities of sonic boom produced by supersonic aircraft are not readily apparent." The report said it could not totally rule out the possibility that unconventional aircraft designs might be devised which could significantly reduce the boom, but it added that the most realistic expectation was for small reductions obtained through better understanding of theory, refinements in the design of conventional aircraft, and improvements in propulsive efficiency and operating procedures.[15] In June 1968 the subcommittee issued a second report that was a shade more optimistic but it remained "unaware of any major breakthroughs in the sonic boom problem."[16]

The initial report was followed by a highly controversial report which brought charges that the Academy was minimizing the damage that the sonic boom would produce. The second report, completed in February 1968, dealt with "physical effects of the sonic boom." Its major conclusion was that "the probability of material damage being caused by sonic booms generated by aircraft operating supersonically in a safe, normal manner is very small."[17] There was later much debate over precisely what this sentence meant, but it was widely interpreted at the time as indicating that one need not fear much physical damage from extensive SST flights over land.

In Cambridge, Massachusetts, the report on physical effects was read with dismay by William A. Shurcliff, a physicist employed at the Cambridge Electron Accelerator who served, in his spare time, as organizer and director of the Citizens League against the Sonic Boom, a loosely organized group of several thousand people concerned with the SST's impact on the environment. Shurcliff, a prolific pamphleteer, had been sending literature to lawmakers, newsmen, and citizens' groups purporting to show that the sonic boom would cause extensive damage to glass, plaster, masonry, and various building elements that might have been poorly constructed or put under stress before being hit with the SST's shock wave. Yet here was the nation's leading scientific body suggesting that the damage would be negligible. The question of damage was a significant one, for it might affect whether the SST would be allowed to fly over populated land areas and, perhaps, whether the SST would be built at all. Upset by what he regarded as inaccuracies, omissions, and half-truths in the Academy report, Shurcliff launched a one-man campaign to get the Academy to issue a correction. The story of his struggle reveals just how difficult it is to get the Academy to admit error—even when a substantial portion of its own members clamor for a retraction.

Shurcliff opened his campaign by writing to John Dunning, chairman of the SST–Sonic Boom Committee, and to the members of the subcommittee that had produced the physical effects report, seeking a correction or clarification. He pointed out that there had been substantial damage payments made in connection with sonic boom tests in St. Louis and Oklahoma City—and yet the press was reporting, on the basis of the Academy report, that the likelihood of sonic boom damage was small. A more accurate statement, Shurcliff suggested, would be that, while the chance that any one specified plane would damage any one specified house was indeed very small, nevertheless "because there would be so many planes, each inflicting its boom on so many houses, the aggregate damage would, in all likelihood, be on the order of $1,000,000 per day."[18] Shurcliff based his estimate on an extrapolation from damage claims data compiled in cities that had already been subjected to sonic boom tests. Shurcliff's estimate was considerably higher than most such calculations, but he was hardly alone in his fears. A high-level scientific advisory group appointed by the Secretary of the Interior later said a "conservative estimate" of the continuing annual cost of repairing sonic boom damage to houses and other structures would be "at least $35 million, and possibly more than $80 million per year."[19]

Shurcliff had no luck in his initial efforts to gain a correction from the committee, so he escalated his campaign by writing to all members of the governing board of the Academy. That, too, produced nothing more than a few vague letters in reply, so Shurcliff finally pulled his trump card. On August 8, 1968, roughly six months after the original report had been completed, the Citizens League sent a letter to all members of the Academy. The letter was signed by Shurcliff and by John T. Edsall, a Harvard biologist who was deputy director of the Citizens League and a respected member of the Academy himself. The letter was accompanied by a fact sheet purporting to prove that the main conclusion statement of the original report was "false," and it concluded:

"Because the National Academy is the leading body which speaks for American science, it is of major importance that the reports which the Academy sponsors shall be scrupulously accurate and objective. The publication of such statements as that quoted above damages the reputation of the Academy and misleads the public. Will you join us in asking the Academy to correct the statement, and if so, would you be willing to sign the enclosed form and return it to us?"[20]

Faced with the threat of an insurrection among its own membership, the Academy leadership finally struck back. On August 22, 1968, a packet of material with a cover letter signed by John S. Coleman, the Academy's executive officer, went out to all Academy members. The letter noted that twelve of the sixteen members of the SST–Sonic Boom Committee were members of either the National Academy of Sciences or the National Academy of Engineering and that "all are distinguished in their respective areas of specialization"—a hint presumably that it would be foolhardy or impolite for Academy members to challenge the judgment of such distinguished peers. The letter also said that the executive committee of the NAS Council had reviewed Shurcliff's charges and "agreed that the meaning of the offending sentence could be construed as Mr. Shurcliff has construed it."[21] The letter failed to note that virtually everyone else who had cited the report—the press and even the Academy's own press release and its internal newsletter—had also construed it the way Shurcliff did: namely, that sonic boom damage would be minimal.

Included in the packet of material was a "clarifying statement" issued by the Committee on SST–Sonic Boom. The thrust of this statement was that all those who had been complaining had simply misread the report:

> They have interpreted a statement made in the report to indicate a belief on the part of the Subcommittee that commercial operations over the continental United States of a large fleet of SST's having characteristics similar to those projected for the Boeing competition design would result in insignificant cumulative damage to glass panes, plaster, or physical structures. The Subcommittee statement which has been challenged was not intended to convey this meaning.[22]

The clarifying statement went on to explain that the report was based on certain crucial assumptions that had perhaps not been made clear. The chief assumption was that "human annoyance in response to booms"—the subject of an Academy report that had not yet been completed—"would provide even greater constraints on SST designs and patterns of overland operations than those imposed by possible material damage." Thus, "to be acceptable to people," an SST would have to be designed so that the overpressure characteristics of its sonic boom were "significantly less in value than those anticipated for the Boeing competition SST design." If such an aircraft with acceptable "people" response qualities were developed, the clarifying statement said, then "the probability of substantial structural damage to *any single structure* [emphasis added] would be very small." The statement said it was "not the Subcommittee's intent to attempt an evaluation of the cumulative damage which might result from one or more transcontinental overflights of a supersonic transport of present or proposed design." It said such an evaluation would be of "questionable accuracy" because of limited data. The only reason the subcommittee had made its original statement about the probability of structural damage being "very small," the clarifying material implied, was to point out that it would be difficult to conduct field tests by exposing a particular house or houses to sonic booms. As a result, controlled experiments would have to be conducted "under laboratory conditions." Tacked on to the end of the clarifying statement was an assertion by the officers of the Academy that "The Academy is deeply aware of many threats to our environment. . . . It has not advocated any decision to proceed with the development of an SST transport for overland civilian service."

The remarkable thing about this clarifying material is that it was considerably stronger than any previous statements the Academy had made. It said that the Boeing design SST—the one that was the focus of the great political debate then raging in the country—would almost certainly produce a sonic boom that would be unacceptably annoying to people. And it said that, even if an aircraft acceptable to people could be built, it was impossible, on the basis of existing knowledge, to predict the cumulative physical damage that would result. This was presumably the most accurate assessment that the Academy could make, for it was the end product of months of debate and it had been carefully reviewed by the Academy's executive committee and other internal groups before being issued. Nothing the Academy had said earlier comes close to this statement in emphatically asserting the unacceptability of the proposed Boeing SST for overland flights. Such an assessment by the nation's most prestigious scientific body could presumably have had great impact on the SST debate had it been made publicly. But the statement was never issued. It was simply sent internally to all members of the Academy.

These internal communications, strongly worded though they were, did not quell the issue, for surprising numbers of Academy members responded to Shurcliff's plea that they join him in demanding a correction. Within a few months, 189 members of the Academy—roughly one-fourth of the membership at that time—had signed petitions contending that the February 1968 report on physical effects "seems at variance with the facts" and should therefore be corrected. Some penned in their own indignant comments or wrote letters to accompany the form statement of the petition.

"Essentially, what I read into the original report, and the recent clarifying statement, is this: If a boomless supersonic plane can be built, its booms will produce negligible physical damage," commented one Academician. "Such a piece of foxy political verbiage has no place in advice given to the government by the NAS," replied another. "I am horrified—it is time the NAS either stop issuing reports on public policy or get more competent people to issue them," added a third. "Certainly this sentence [the main conclusion of physical effects report] will be used by anyone who favors SST to 'prove' that booms are unimportant—it must be clarified," complained a fourth. "This appears to be the result of the technological, as opposed to ecological, bias built into the committee by the Governing Body of the Academy," said a fifth.[23]

This surprising flood of criticism—perhaps unprecedented in Academy history—would seem to indicate that something was very wrong with the report. But Dunning, the chairman of the parent committee, believes the 189 Academicians who signed petitions criticizing the report were acting more from emotion than from reason. "The plain truth is that scientists, Nobel Prize winners, engineers and so on get carried away by things as much as anybody else, maybe even more so," he said in an interview.

Nevertheless Shurcliff kept trying to win a public correction. He threatened to send a new batch of material around to Academy members attacking the clarifying material. And he prepared a press release detailing the story of his long battle with the Academy, complete with numerous quotes from Academy members who had responded to his petition. In an effort to head off the bad publicity that would result from issuance of the press release, Merle A. Tuve, home secretary of the Academy, wrote a letter to Shurcliff in which he asserted that Shurcliff had misread the crucial sentence in the original report on physical effects and that the Academy had gone about as far as it could go in overruling its own committee.

At about the same time Shurcliff also got calls from other high-ranking Academy members who were themselves opposed to the SST but who urged him to drop his plans for a press release on the ground, he recalls, that "You've made your point—we know we're wrong but we don't know what to do about it." The protective walls of the scientific fraternity were thus closing around the Academy to shield the revered institution from possible embarrassment. So Shurcliff did drop his plans. And finally, in February 1969, a full year after the original physical effects report had been issued, he received the retraction he had been seeking—sort of. A three-paragraph item tucked away in the back pages of the Academy's newsletter apologized for the fact that "some confusion" had been generated by a news article on the physical effects report a year earlier. The editors, in a statement presumably dictated by higher authority, accepted full blame for the confusion and exonerated the authors of the original report from any intention to minimize sonic boom damage. As the editors expressed it:

"It is quite evident that sonic booms of the intensity that can now be expected from the first generation of supersonic aircraft pose real problems in terms of human annoyance; experience has also shown that some property damage can be anticipated were such planes to fly over populated areas."[24]

So far as can be determined, virtually no one who was not deeply involved in the long internal battle took much notice of the clarification. Indeed, writers and public figures continue to refer to the original report, and Shurcliff has been kept busy chasing down the misconceptions. In June 1970, Shurcliff discovered that Ralph Tyler Smith, then a Republican Senator from Illinois, was citing the Academy's sonic boom studies as evidence that misgivings about the SST were "unsubstantiated."[25] At Shurcliff's request, the Academy contacted Smith's staff and secured an agreement that the Senator would no longer distribute the statement he had been making.[26] Shortly afterward, Shurcliff discovered that a forthcoming book on noise pollution carried the controversial statement from the physical effects report. At his request, the president of the Academy wrote to the publisher that the original statement "was subsequently reconsidered by an appropriate committee of this Academy and was retracted."[27] Just what practical difference such retraction makes is dubious. Today, if a member of the public asks for the Academy's sonic boom reports, he is given the original reports but not the clarifying material that was circulated internally.

It has never been clear throughout the whole long struggle just what the sonic boom committee really did mean in its original report on physical effects. Was the committee simply guilty of sloppy writing which gave others the impression that structural damage would be no problem? Or did it really believe that such damage would be minimal and only fall back on the "we've been misunderstood" defense when the heat became too great? The report is so poorly written that one can read it either way, and there is fragmentary evidence, in the form of internal letters and early reports, to support either interpretation. Perhaps the committee was of mixed mind or else had no position on the issue. But it would seem significant that the Academy press release which announced publication of the report was cleared with someone in authority on the committee, and it stated emphatically, in its lead sentence, that "the probability of damage to structures from sonic booms generated by aircraft operating supersonically in a normal manner is small."[28]

The sonic boom committee's third report, dealing with "human response to the sonic boom," was issued in June 1968, right in the midst of Shurcliff's battle to get the physical effects report corrected. The "human response" report was sharply criticized by Shurcliff, too, not so much for what it said as for what it failed to say. The report was little more than a list of recommendations for research to determine the impact of sonic booms on communities and individuals, including such sensitive groups as infants, the elderly, and hospital populations; but it shed little light on whether the sonic boom would actually prove harmful or annoying to humans.

The report made no effort to evaluate the significance of past research on human response, explaining that such information was "readily available in the public domain" and could be obtained from various government agencies.[29] It did not try to estimate what range of boom intensity would be tolerable or intolerable because, as the chairman later explained, "The research done to date is quite inadequate for any reasonable estimate as to what might be an acceptable boom level."[30] It made no attempt to assess the numbers of people who would be disturbed by the boom. And, on the crucial political issue of whether the first generation of SST's would be able to fly over land, it was vague and timorous; it assumed at two points in the text that the first generation SST's would probably be restricted to routes over water, but it did not recommend such restrictions.[31] A cover letter attached by Raymond A. Bauer, professor of business administration at Harvard, was a trifle stronger, but not much. It said the boom was not apt to damage hearing or cause other direct physiological harm, but that "a review of field studies of the psychological impact of the sonic boom shows a growing consensus that is discouraging for the use of the current version of the commercial supersonic transport (SST) over populated areas at speeds at which it will be generating a sonic boom." After reading the report, Shurcliff complained: "The people of this country have been waiting for years to receive some reliable statement on the human response to the sonic boom. The long-heralded report comes out—and leaves the people empty handed. The main conclusions presented ... are couched in such vague language that they carry no clear message."[32]

The sonic boom committee sputtered on for another three years after issuing its public reports. It published an annotated bibliography on animal response to sonic booms in 1970, submitted a final report (unpublished) in June 1971, and finally closed down operations effective June 30, 1971. The committee's seven-year effort had undoubtedly helped the government plan a research program to test and analyze the impact of the boom. According to Russell Drew, who handled SST matters for the now-defunct White House Office of Science and Technology, the Academy's analysis was "as valuable as any around—it was used and played a role." But the Academy's impact on the policy aspects of the SST debate—such as whether the plane should be built or should be allowed to fly over land—was negligible. Whatever comfort SST proponents gleaned from the Academy's controversial report on physical damage disappeared when the SST ultimately went down to defeat anyway in the Congress. Moreover, even before the SST's defeat, the whole issue of sonic boom effects became largely moot when the government, under great political pressure, announced that it would ban supersonic flights over populated areas.

The most significant aspect of the sonic boom committee's seven-year history, from the viewpoint of this present study, was its revelation of serious flaws in the Academy's relationship to its governmental funding agency—flaws which hampered the Academy's ability to grapple with a controversial issue. To begin with, the Academy had allowed itself to be used by the government to head off public opposition to the Oklahoma City tests. It had not been asked, nor had it sought, to play a role in the sonic boom problem earlier. Once involved in the project, the Academy had shown a disturbing tendency to line itself up with the sponsoring agency, the FAA, against the agency's critics. This became apparent when the Academy made itself a public relations adviser to the government and when it declined to make any strong statements against the SST even though internal documents indicate that the committee felt the Boeing SST would be unacceptably annoying to the public. Then, when a strong and apparently legitimate challenge was made of a committee report, the Academy had shown an inability to admit error. Whether one thinks that the committee erred on matters of substance or merely wrote a report that was subject to misunderstanding, the Academy had a responsibility to make its position clear. It did so in internal documents but failed to notify the public in any effective way.

Finally, and perhaps most significant of all, the Academy allowed itself to be put in a straitjacket by accepting an overly narrow definition of its task from the government. The committee seems at first to have interpreted its mandate rather broadly. Its first two status reports said the committee would, among other things, conduct an "examination and analysis of available data on the sonic boom for the purpose of assisting in determining the feasibility of SST operations."[33] But in later years, as the committee prepared to issue its first public reports, it narrowed its focus and pretty much limited itself to recommending research that might help shed light on the sonic boom phenomenon. The justification given for this approach was that a memorandum signed by President Johnson on May 20, 1964—the day the FAA announced that it had called in the Academy to review the Oklahoma City tests—had requested that the Academy "plan an expanded sonic boom program, specify the tests that may be desirable, monitor the program, and analyze the data derived from the tests...."[34] Colonel John P. Taylor, the retired air force officer who staffed the committee during most of its existence, recalls that he would cite the memorandum at virtually every meeting and remind the committee that it was only to recommend research. This approach effectively precluded any statement that might interfere with the government's plans to produce an SST.

Some opponents of the SST regard the Academy as an apologist for the SST, but the truth seems to be that Academy committees were probably inclined to make even stronger statements against the sonic boom than they were able to within the confines of their task. This became evident when the chairmen of two subcommittees made statements, as private individuals, that went considerably beyond what they had felt able to say in Academy reports. Thus Raymond Bauer, head of the group that put out the "human response" report, proclaimed in a speech shortly afterward that "the prospect of an SST which is both economically viable and socially acceptable is quite distant. This area of technology is well beyond my competence, but everything that I read and hear suggests that without some new breakthrough the solution is remote." Bauer also said that SST flights over land "would involve the vast majority of American citizens and communities," and that the booms generated by the Concorde and Boeing SST's would be more than twice the level he provisionally defined as acceptable.[35] Those were strong statements. Yet the report of Bauer's subcommittee had largely limited itself to recommending research and only in a single cover letter had it even hinted that the outlook for an acceptable SST was "discouraging." Similarly, Raymond L. Bisplinghoff, head of the group that put out the Academy report on sonic boom generation, urged in Congressional testimony in 1970 that a statement be made that "under no circumstances" could the Boeing design SST "ever be operated over populated areas" (emphasis his).[36] Yet the report of Bisplinghoff's committee limits itself to recommending research and makes no reference to the implications of sonic boom for overland flights. Unfortunately, by limiting itself to the innocuous role of recommending a research program, the Academy effectively removed itself from meaningful impact on one of the major technological issues of recent years.

The Academy also considered a second potential environmental impact of the SST—the possibility that the SST exhausts might change the composition of the upper atmosphere and thus cause drastic climatic changes that could adversely affect the health or well-being of humans. In 1966 an Academy panel on weather and climate modification issued a two-volume report which considered, in a brief discussion, whether the release of large volumes of water vapor into the atmosphere from fuel combustion by a fleet of SST's might inadvertently cause harmful effects. The answer was a reassuring no. The two effects then considered most likely—a significant increase in cirrus clouds or a significant increase in the relative humidity of the stratosphere, both of which could affect the radiation balance and possibly the general atmospheric circulation—were judged unlikely to cause appreciable climatic disturbance. As the report phrased it: "Our tentative conclusion, based on an assumed traffic volume of four flights per day for 400 supersonic transport airplanes, is that neither additional cloudiness (contrails) nor water vapor absorption of long-wave radiation will be sufficient to disturb appreciably either stratospheric properties or the large-scale circulations that are influenced by its thermodynamic state."[37]

That conclusion was cited repeatedly by the FAA and other SST proponents to buttress their contention that the SST would not do unwanted environmental damage. Thomas F. Malone, chairman of the panel in later years, complains that the statement was "lifted out of context and used as a reason for not taking a very serious look at some of the atmospheric side effects of the SST," which the original report had not considered. So in 1970 the panel decided, in the course of updating its 1966 report, that it had better take another look at the SST's possible environmental impact.

The scientist the Academy panel appointed to take the new look was the late James E. McDonald, a respected atmospheric researcher from the University of Arizona, who had been involved in the earlier 1966 study. McDonald read deeply in the relevant literature, talked with leading experts, and became increasingly uneasy that the SST might indeed cause an environmental disaster. He concluded that the 1966 report had been right when it discounted the likelihood of contrail formation, but he found new evidence which convinced him that SST exhausts might react unfavorably with the upper atmosphere. He was particularly worried that water vapor exhausts might deplete the ozone shield which screens out ultraviolet radiation from the sun. McDonald's reasoning, as subsequently presented in Congressional testimony, involved four interrelated steps. He warned that (1) the SST would put vast amounts of water vapor into the stratosphere, (2) this water vapor would react with ozone and deplete some fraction of it, (3) depletion of the ozone shield would allow additional amounts of ultraviolet radiation to penetrate to the earth's surface, and (4) this increase in ultraviolet radiation could lead to an increase in skin cancer. In a statement which caused considerable debate in the scientific community, McDonald then went on to predict that full-scale SST operations would produce a 1 percent reduction in ozone in the stratosphere, which in turn could lead to an additional ten thousand skin cancer cases in the United States each year.[38]

Laymen who hear this line of reasoning react with incredulity at first—it seems so far-fetched that planes flying far overhead can cause an increase in skin cancer on the ground. Yet McDonald's paper was reviewed at several levels in the Academy after Malone brought it to the attention of Academy President Handler as something that "should be looked at" even though it might well be "a wild scheme." Since the problem involved medical issues far beyond the expertise of the weather panel, the Academy's division of medical sciences assembled a team of experts for a one-day meeting to consider McDonald's thesis. The upshot of that meeting, according to Charles L. Dunham, division chairman, was agreement that, if there were an increase in ultraviolet exposure as predicted by McDonald, then there would indeed be an increase in the incidence of skin cancer, though the quantitative relations remained in doubt. The Academy made no public statement on the issue, but a majority of the experts assembled by the Academy subsequently made individual statements warning that the skin cancer danger might indeed be real.* McDonald's paper was also reviewed

---

*Three of the four medical experts assembled by the Academy were also among forty experts whose views were sought and made public by Senator William Proxmire (D-Wisc.). All three supported McDonald's reasoning as to the medical implications of increased ultraviolet radiation, though not all of them would necessarily agree that the SST program should therefore

by the Academy's Committee on Science and Public Policy, the most important and prestigious committee in the entire Academy hierarchy and one that is specifically designated to deal with important policy issues. According to Malone, COSPUP was asked, in essence, "Are these concerns legitimate or is it a bunch of horseshit?" and COSPUP replied, "It's not horseshit. We don't know whether it's right or not, but there are no obvious holes in it. Follow it up."

Unfortunately, the way in which the Academy chose to express its concern tended to play into the hands of the pro-SST forces in the Administration. A three-man delegation, consisting of Malone, McDonald, and staff man John Sievers, paid a personal visit to Department of Transportation (DOT) officials in charge of the SST project on November 30, 1970, and communicated McDonald's fears in private. Malone recalls that DOT officials reacted with "incredulity, the same as I had originally expressed. It takes a while to sink in."

The visit caused understandable consternation at DOT. A crucial Senate vote on the SST project was scheduled to take place within a few days, and the skin cancer story was almost certain to become public knowledge, for a reporter

---

be stopped. William L. Epstein, of the University of California Medical School in San Francisco, said: "The problem posed by Dr. McDonald is not merely academic but a significant medical one . . . which should be carefully considered during the early planning stages of the SST and not as an after-the-fact national health problem." Frederick Urbach, of Temple University Skin and Cancer Hospital, stated: "Our best estimate suggests that a decrease of 1 percent in the ozone layer might result some years later in an increase in skin cancer of the order of 10 to 15,000 new cases per year in the United States. This is in keeping with the estimates of Dr. McDonald." Farrington Daniels, Jr., of Cornell Medical Center, carried McDonald's reasoning far beyond the skin cancer issue and warned that since all familiar plants and animals have evolved under the protection of the ozone layer, tampering with that protective layer "might be considered as potentially serious as tampering with the concentration of salt in the sea." The fourth expert who attended the Academy's review session does not seem to have been on the list of experts Proxmire consulted. A fifth expert had also been invited to attend the Academy's review session but had been unable to attend. He was Kendric C. Smith, of Stanford University School of Medicine, and he later told Proxmire: "Dr. McDonald's conclusions appear quite reasonable to me and to other scientists that I have talked to who have seen his preliminary report. When he makes a calculation he uses the lowest estimate of the probability in order to err on the conservative side." The statements by Urbach, Daniels and Smith can be found in the *Congressional Record*, vol. 117, pt. 6, March 19, 1971, 7258, 7261–7263. The letter from Epstein is in Proxmire's files.

was at DOT headquarters waiting to intercept the Academy delegation. So DOT prevailed upon the Academy to disown McDonald. Reporters and Congressional staffers who telephoned the Academy about the skin cancer issue during the next few days were told that the theory represented McDonald's thinking, not the Academy's. Senator Barry Goldwater (R-Ariz.), for example, stressed in floor debate that McDonald's theory had been released "without appropriate Academy concurrence or review, even though this is part of an Academy study on climate modification."[39] And when McDonald later was scheduled to testify before a House committee on March 2, 1971, DOT had the Academy disown the testimony the day before it was given. On March 1, Malone sent a telegram to William Magruder, director of SST development at DOT, advising him that the Academy had made no statement "that the SST could increase the incidence of skin cancer." Malone added that the findings of the weather modification panel would only be released after "the normal NAS procedure of a thorough scientific review to insure their validity."[40]

Malone's own views were forcefully expressed in a letter to a senator two weeks later. Malone described McDonald's conclusions on ozone depletion as "fraught with uncertainty" and he said the evidence was "most tenuous" concerning a likely increase in the incidence of skin cancer. "McDonald's testimony has substantial usefulness in identifying questions that merit study but as a document that might be useful in resolving the public concern over the environmental effects of large-scale SST operations, it is much less than persuasive," Malone wrote. "Nor was it ever the intent of our panel to develop an authoritative treatment that would be useful in establishing public policy. Our objective was simply to identify some meaningful questions and bring them to the attention of photochemists, photobiologists and others who could bring to bear the mature insight that years of study leave with a scientist."[41]

The effect of this posture was to minimize the Academy's apparently real concern that the SST might cause severe atmospheric problems. In the beginning, Academy officials apparently felt that McDonald might be on the right track even though his theory was far from proven, so they encouraged him to communicate his concerns to DOT, and later to a

Commerce Department group. But when the skin cancer theory became public knowledge, the Academy quickly back-pedaled and allowed itself to be used in a way which tended to discredit McDonald. This left McDonald as a lone voice crying alarm, and unfortunately his was a voice that was easy to discredit. Although McDonald (who died in June 1971, an apparent suicide)[42] was a respected atmospheric scientist, he had advanced some controversial theories regarding unidentified flying objects. It was thus not difficult to picture the bizarre skin cancer theory as but the latest invention of a flying-saucer nut. And that's exactly what happened. The Associated Press reported that there was "stunned silence" during an SST hearing when McDonald was questioned about his belief that unidentified flying objects were connected with the great New York power blackout.[43] Later, Vice President Spiro T. Agnew put McDonald "solidly in the ranks of those English doctors who objected to smallpox vaccinations on the theory that they would make people look like cows."[44]

The SST was eventually defeated by a close vote in the Senate in early 1971. Whether McDonald's theory swayed any votes is difficult to determine. But the Academy's unwillingness to say anything publicly about McDonald's theory left both sides of the SST debate feeling cheated.

The reason the Academy backed off seems to be that it didn't want to be accused of undermining the SST program on the basis of insufficient information. "Until I felt more sure that we were on the right track I was reluctant to surface this thing," Malone recalls.

The Academy subsequently set up study groups to study the issues raised by McDonald in greater depth. The question remained important because foreign-built SST's will soon be circling the globe, and the American SST project may someday be revived.

Early reports from these study groups concluded that there was indeed a possibility, though not a certainty, that SST exhausts would deplete the ozone and thus cause biological damage. But the focus of concern shifted. Whereas McDonald had suggested that water vapor was the chief concern, subsequent investigators suggested that nitric oxides were actually the chief danger. The leading proponent of this view, Harold S. Johnston, professor of chemistry at the University of California's Berkeley campus, predicted that nitric oxides from a fleet of five hundred SST's might deplete the ozone shield to such an extent that all animals living out of doors during the daytime would be blinded by ultraviolet radiation.[45*]

The Academy convened a panel of atmospheric scientists and chemists on July 29, 1971, to review Johnston's theory of the chemical processes involved as well as critiques of Johnston's theory by a score of scientists. The panel expressed reservations about the adequacy of Johnston's calculations, but there was "general agreement" that his conclusions were "credible ... and that the possibility of serious effects on the normal ozone content cannot be dismissed."[46]

Meanwhile, a panel of medical and biological experts assembled by the Academy's Environmental Studies Board concluded in a 1973 report that, if the SST does result in an increase in ultraviolet radiation reaching the earth's surface, then there is a high probability that human and other life systems would suffer significant harm. Using "conservative" assumptions, the panel estimated that a 5 percent reduction in the ozone shield would produce at least eight thousand additional cases of skin cancer per year in the white population of the United States, of which about three hundred cases would lead to death.[47] Even more significant, according to the panel, was the possibility that increased ultraviolet radiation might diminish the biological productivity of the ocean, interfere with mating and other behavioral patterns of insects and lower animals, and damage plants, especially agricultural species. "Sufficient knowledge is at hand to warrant utmost concern over the possible detrimental effects on our environment by the operations of large numbers of supersonic aircraft," the panel said.[48] That warning was the strongest ever issued against the SST by the Academy. It may even have been too strong, since the question of whether the SST's would, in fact, damage the ozone shield was still a topic of scientific debate. But the

---

*According to Malone, McDonald was also worried that nitrogen oxides in the SST exhausts might deplete the ozone, but, in the hectic scramble to document his concerns over a potential atmospheric catastrophe, he first completed the calculations for water vapor and had not yet gone on to develop the case for nitric oxides.

Academy had finally found the courage to describe the damaging biological consequences should the ozone shield be depleted. Much the same warning should have been issued when McDonald's theory first surfaced.

A more definitive answer to the possible atmospheric and biological impact of the SST was being sought in 1973–1974 by the Department of Transportation's Climatic Impact Assessment Program, with the advisory help of yet another Academy committee. Preliminary results indicated that nitrogen oxides would indeed deplete the ozone shield but that this danger could be avoided by modifying the SST engines so as to reduce nitrogen oxide emissions. If necessary, the size of the SST fleet and the scope of its operations could also be restricted. Water vapor emissions were not judged a serious problem.[49] Thus the chemistry initially emphasized by McDonald turned out to be wrong, but the potential catastrophe he warned against—depletion of the ozone shield—was apparently real. McDonald clearly performed a major public service by pointing to atmospheric problems that had not even been considered by the hundreds of engineers and scientists in industry and government who were rushing to complete the SST. In retrospect, the Academy's failure to endorse his concerns at the time of the SST debate seems an act of cowardice.

One noted Academician who attended some of the meetings where McDonald's theory was discussed believes the Academy should be given credit for proceeding cautiously on an issue "newly opened and totally unexplored," particularly since, "as it turned out, the chemistry initially suggested was wrong" and McDonald's UFO advocacy "did not lend credibility."

The Academy was unquestionably in a tight spot when confronted with McDonald's thesis, for it is traditionally reluctant to make pronouncements until sufficient data is in hand. Yet if the issue was important enough to bring to DOT's attention—as the Academy clearly believed—then it was presumably important enough to bring to the attention of everyone concerned with the SST debate. Instead, the Academy chose to let a single man pretty much present the thesis as an individual, and it had him present the evidence to the two agencies—DOT and later Commerce—which were most enthusiastic about the SST and thus would tend to belittle the theory. Meanwhile, opponents of the SST in Congress at first had difficulty obtaining copies of McDonald's paper, though he was later called to testify.

The Academy's behavior may reflect scientific caution, or a reluctance to place itself in an adversary position toward the executive branch (particularly since the data was less than conclusive), or a preference for dealing with the executive branch rather than Congress. The Academy was obviously worried that it might be subjected to criticism if it voiced fears about the SST without strong documentation. But the Academy seems not to have considered that a failure to voice genuine concerns would also exert influence on the debate, albeit in the opposite, or pro-SST, direction. At the very least the Academy could have made a statement to the effect that McDonald had raised valid questions, and then let the contending political forces fight it out as to whether the SST program should be stopped or should continue while the answers to these questions were sought.

The Academy's failure to speak out clearly on the SST issue is distressing in view of the fate that befell two other SST reports prepared by high-level advisory groups. In early 1969, as the SST controversy was heating up, President Nixon appointed an SST Ad Hoc Review Committee composed of undersecretaries and other key officials from eleven federal agencies. This group, which was chaired by James M. Beggs, Undersecretary of Transportation, produced a report in March 1969 that was largely unfavorable to the SST. It cast doubt on the economic viability of the plane, predicted that the SST would have an adverse impact on balance of payments, and raised disturbing environmental questions. The committee's environmental panel, consisting of high officials from the Office of Science and Technology, the Department of Health, Education, and Welfare, and the Department of the Interior, warned that the sonic boom would be "considered intolerable by a very high percentage of the people affected," a circumstance which made it "essential" that supersonic flights over land be prohibited. The panel also warned that, while atmospheric effects would probably be minor, they "certainly should not be neglected"[50] (this judgment was made before the skin cancer theory had been advanced). A particularly strong anti-

SST statement was made by Lee A. DuBridge, the president's science adviser, in a separate letter to Beggs. DuBridge questioned the commercial viability of the SST, called the noise problem "a matter of worry," described the sonic boom problem as "unsolved," called for "a policy statement that there shall be no supersonic operations by the SST over any populated areas," and questioned whether the SST would achieve the payload and range desired. "Granted that this is an exciting technological development," he said, "it still seems best to me to avoid the serious environmental and nuisance problems and the Government should not be subsidizing a device which has neither commercial attractiveness nor public acceptance."[51]

An equally strong stand against the SST was taken by a special committee of eminent scientists appointed to review the program by the White House Office of Science and Technology. This panel, which was headed by Richard L. Garwin, an IBM physicist, actually recommended that the government withdraw its support from the development of SST prototypes. The group based its conclusion largely on economic grounds, but it also cited possible environmental hazards and pointed out that the environmental considerations would affect the economic viability of the plane. Among adverse effects on the environment, the committee listed noise in the vicinity of airports and "possible influence on the climate."[52] The committee was particularly outspoken on the sonic boom issue. It urged the government to state the "SST's producing a boom intensity in excess of one pound per square foot can clearly not be operated acceptably over land, that all presently conceived SST's far exceed this intensity, and thus will without question be denied operating permission over the U.S."[53]

These two reports—the Beggs and Garwin committee reviews—are strikingly more forceful than the Academy's reports, particularly on the sonic boom issue. But their impact was weakened by a circumstance that did not afflict the Academy reports. Both reports had been prepared by groups operating within the Nixon Administration, and the administration was therefore able to suppress them. The Beggs committee review was prepared in March 1969 but was kept under wraps until October 31, 1969, when Representative Sidney R. Yates (D-Ill.) finally managed to obtain a copy and made it public. Similarly, the Garwin committee report, which had been submitted on March 30, 1969, was not made public until August 21, 1971, long after the SST project was dead—and then it had taken a court suit to force the government's hand. In the interim, the Nixon Administration whipped its scientific troops into line and had its top scientists come out in favor of the SST. Although DuBridge, in connection with the suppressed Beggs report, had opposed the SST, he subsequently adopted the Administration line, explaining that "The President has a broader view of the whole problem."[54] And while the Administration sat on the hostile Garwin report, Edward David, Jr., who succeeded DuBridge as science adviser, solicited pro-SST testimonials from scientists who could be persuaded to support the program.[55]

Such shenanigans are not unusual in Washington. There is always a danger that administration-appointed committees will have their reports ignored or bottled up if the findings are at variance with administration policy. Some will even argue that this is as it should be. But the mere fact that such suppression occurs would seem to make it imperative that the supposedly independent Academy speak out clearly and unequivocally. Its failure to do so robbed Congress and the country of an objective, authoritative, unfettered scientific voice.

CHAPTER SEVEN

Defoliation:

The Academy as Shield for the Pentagon

"We were getting the endorsement of the gods."

—Donald M. MacArthur, former Deputy Director of Defense Research and Engineering, in an interview explaining why the Department of Defense asked the Academy to participate in an assessment of the military herbicide program, July 13, 1971.

T he story behind the Academy's involvement in the controversy over military use of herbicides in Vietnam illustrates an important fact about Washington politics: investigations are not always what they seem. Twice in recent years the Academy has participated in studies aimed at determining whether the herbicide program has inadvertently caused an ecological or medical catastrophe. In 1967, the Academy was asked to review a survey of scientific articles relating to herbicides; and in 1970, the Academy was asked to launch an actual field study in South Vietnam. Ordinarily, one might expect that military authorities would be apprehensive at the prospect of such scrutiny by supposedly objective outsiders. But such was not the case in these two instances. The Department of Defense (DOD) actually embraced the Academy investigators as allies who would help support the herbicide program against criticism. The Pentagon used the prestige of the Academy as a weapon to beat back challenges from other scientists; it maneuvered the Academy into positions where it could do little to undercut the herbicide program until after it had been phased out; and it even made plans to take the results of the Academy field study and apply them to future military planning. In much of this, the Academy seems to have been an unwitting pawn of the Pentagon, but there is also evidence that the Academy, in some instances, was a willing victim. Whatever the case may be, there is little doubt that the Academy failed to act as a vigorous, impartial overseer of the herbicide program. Instead, it ended up serving as a shield for the program, while another scientific organization—the normally somnolent American Association for the Advancement of Science (AAAS)—reluctantly had to assume a watchdog role over the program.*

---

*Since this book argues for full disclosure of all possible bias, it should be pointed out that Philip M. Boffey, the author of this work, was formerly employed by the AAAS as staff writer on the association's journal, *Science*, from 1967 to 1971.

The military herbicide program started modestly in late 1961 when experiments were conducted in Vietnam to determine whether chemical agents sprayed from airplanes could be used to strip foliage from trees (defoliation) or to kill food crops. The early tests were judged a success. Two chemicals—known as 2,4-D and 2,4,5-T—were found to be effective defoliants, and a third chemical—cacodylic acid—showed promise for use against crops. In 1962, the herbicide program became operational. Low-flying C-123 cargo aircraft, equipped with special tanks and nozzles, began to make regular spray missions. At first the number of missions flown and the amount of herbicide dropped was relatively small, but by late 1965 the program began to expand enormously, finally reaching a peak in 1967 before declining in subsequent years. Over a six-year period, between 1965 and 1971, about one-twelfth of the land area of South Vietnam—an area larger than the state of Connecticut—was treated with herbicides. Chemical defoliants were dropped along the edges of rivers, canals, roads, and railroads in an effort to prevent ambushes; around villages and military posts in an effort to prevent sneak attacks; along enemy supply routes; in the Demilitarized Zone; and over great swatches of forest to create bare stretches so that enemy movements might be observed from the air. Anticrop agents were dropped on fields of manioc, sweet potatoes, rice, vegetables and other foodstuffs believed destined for enemy mouths.[1]

In the early stages of the program military scientists seem to have given little thought to possible long-range effects of herbicides on the ecology.[2] But as the program expanded and became better known in the civilian scientific community, strong protests were lodged by ad hoc groups of eminent scientists. Some attacked the anticrop chemicals as indiscriminate in their choice of victims, harming civilians as well as soldiers. Others warned that widespread defoliation might cause lasting ecological damage or subtle long-term medical problems. And still others deplored any use of chemical herbicides lest the door then be opened to use yet more lethal chemicals in warfare. Although the National Academy of Sciences played no role as an institution in any of these early protests, many of its members participated as individuals.

The first significant effort to get a major institution of the scientific community involved in questioning the defoliation program took place not in the Academy, with its close ties to government, but in the American Association for the Advancement of Science (AAAS), an independent organization which is closer to "grass roots" opinion in the scientific community than the more elite Academy. The issue was first brought to AAAS attention in 1966 by E. W. Pfeiffer, a zoologist at the University of Montana, who urged the AAAS to sponsor a scientific field investigation of chemical and biological warfare (CBW) effects in Vietnam.[3]

Much to the consternation of Pfeiffer, the initial reaction of the AAAS was to try to refer the problem to the Academy, largely because the Academy had greater experience in conducting large studies and in dealing with the government. Pfeiffer dissented strenuously. He argued that the AAAS should do the study itself because the Academy was too closely associated with Fort Detrick, the Army's biological warfare research center in Frederick, Maryland. "I believe that a truly independent organization of scientists should make this study," Pfeiffer said. "I do not believe that the National Academy of Sciences is such an organization."[4] As evidence to support his view, he cited an article in *Science*, the AAAS journal, which described the Academy as a "source of advice for the biological warfare effort."[5]

The Academy did indeed have a history of involvement in the CBW effort. It had advised the government on gas warfare problems in World War I, and it had stimulated and supervised the biological warfare effort in World War II.[6] It had even helped to develop herbicides for military use in crop destruction. During World War II, an Academy committee evaluated new research on the use of growth-regulating substances for crop destruction and submitted a report on it to the War Research Service, with the result that the Army launched an herbicide research program at Camp Detrick (later renamed Fort Detrick), Maryland.[7] Academy personnel went far beyond their usual advisory role in many of these biological warfare tasks; according to an Academy report they exercised "top supervision of the entire undertaking."[8] The details are still classified, but Frederick Seitz, former president of the Academy, recalls that the Academy was "in the thick of both biological and chemical warfare—it was actually conducting experiments." He adds: "During the war you did what you thought was in the national interest."

After the war the Academy continued to assist the **CBW** program by reviewing a classified report on biological warfare in 1950,[9] sponsoring a conference on airborne infection that was partly supported by Fort Detrick in 1960, and administering, from 1958 to 1970, a fellowship program that was designed to help bring scientific talent into Detrick.[10]

Thus the AAAS was turning for an impartial assessment to an institution which had been instrumental in developing the very defoliants which were now under attack for producing damaging side effects. Whether such previous involvement in the herbicides program necessarily made it impossible for the Academy to conduct an objective study of one aspect of that program—the spraying in Vietnam—is debatable. The Academy was not, after all, serving as an adviser to the Vietnam spray campaign. But to the extent that there was an institutional bias at the Academy, it would tend to support a program that the Academy had helped to foster.

These considerations do not seem to have troubled the AAAS leadership. On September 13, 1967, the president of the AAAS wrote to the Secretary of Defense urging that a study by an independent scientific institution or committee be undertaken. The letter said the AAAS was not "equipped to conduct such a study" and suggested that the Academy would be an appropriate institution, with some kind of independent commission being an acceptable alternative.[11]

Neither the Pentagon nor the Academy seemed interested in conducting the kind of field study the AAAS had in mind. Instead, the Pentagon orchestrated an alternative that was designed to deflect the AAAS push without really answering the questions that had been raised. To begin with, Defense science officials tried to create the impression that they were already looking into the problem without prodding from the AAAS. When AAAS officials first conferred with Pentagon scientists, in late July 1967, they presented a draft of the letter they intended to send requesting that a field study be organized. The Pentagon officials asked the AAAS to delay sending the letter for two months—a request that was granted. Then, when the letter was finally sent, on September 13, the Pentagon was able to reply that a contractor was already at work on an assessment of the problem.[12] John S. Foster, Jr., director of defense research and engineering, assured the AAAS that "qualified scientists, both inside and outside our government," had judged that "seriously adverse consequences" would not occur from the spray campaign. However, in order to allay any remaining fears, Foster said the Defense Department had "recently commissioned a leading non-profit research institute to thoroughly review and assess all current data in this area" and had "recently asked Dr. Frederick Seitz to assemble a group within the National Academy of Sciences–National Research Council to review the results of the study and to make appropriate recommendations concerning it."[13]

The contractor chosen to perform the assessment was Midwest Research Institute (MRI) of Kansas City, Missouri, an organization which had previously worked on CBW matters for the Defense Department but which had not, so far as is publicly known, been involved in the development of herbicides for military use. The MRI's job was to review the published, unclassified literature relating to ecological consequences of repeated or extensive use of herbicides, particularly those herbicides being used in Vietnam, and to consult knowledgeable individuals and institutions.[14] The report prepared by MRI was to be sent to the Academy for review, and the whole batch of material would then be made available to AAAS.

The Academy had thus been brought into the controversy, but not in the role the AAAS had sought. The Academy was not going to do a field study—there was not going to be any field study at all. (Pentagon officials consistently argued that no field study could be safely conducted while the war was on, but critics in the scientific community considered this a half-truth and evasion.) The Academy was not even going to perform the literature review itself. That job had been given to a contractor. The Academy's sole role would be to review the report prepared by Midwest Research Institute and say whether it thought the library work had been competently done. The role was so narrowly defined that the Academy was not even supposed to say whether it agreed with the MRI report or not and was not supposed to even discuss the questions that had been raised by AAAS. As Seitz later described the mission:

> The review would constitute an assessment of the thoroughness and accuracy with which the scientific literature relating to herbicides and their ecological effects had been examined

and evaluated by the Institute. The reviewers were not asked to consider the specific issue of how well or how fully the report responds to the questions expressed in the American Association for the Advancement of Science resolution of December, 1966. Also, they were not asked to endorse, approve, or reject the report.[15]

The review role that the Academy accepted was probably the minimal response that could be made without turning DOD down completely, but it had the unfortunate result of putting the Academy in a position where it could say little or nothing that would embarrass DOD, yet its prestige could be used by DOD to blunt criticism of the program. "We were getting the endorsement of the gods," explains Donald M. MacArthur, then deputy director of defense research and engineering. "If the Academy had said the MRI Report was lousy and torn it to bits, we'd have been in hot water. But if the Academy review came out well it could help us. We did not know the outcome, of course, but we were pretty sure that MRI would do a good job."

The odds were rather good, as it turned out, that the Academy review would not be too damaging to DOD—for the project was placed in hands "friendly" to DOD from the start. To begin with, there was a conflict of interest at the very highest levels of the Academy. Frederick Seitz, then the Academy president, was simultaneously serving as chairman of the Defense Science Board, the Pentagon's highest science advisory body, a fact which would make it difficult for the Academy to criticize DOD under the best of circumstances. But, to make matters worse, the administration of this particular review was placed in the hands of A. Geoffrey Norman, vice-president for research at the University of Michigan, who had served as a biochemist and division chief at Fort Detrick from 1946 to 1952 and in that capacity had become one of the most influential figures in the Army's herbicide research program. What's more, Norman had directed the synthesis and testing of nearly 1100 substances in an herbicide research program at Detrick in 1944–1945.[16] Thus Norman was being asked to quarterback a supposedly objective evaluation of the very program he had helped father. Norman recalls that he was sensitive to the problem of conflict of interest and that he therefore excluded himself from the scientific review panel. "Because of my prior involvement, much earlier on, with the Chemical Corps," he says, "I stayed out of the scientific evaluation entirely." Nevertheless, Norman did play a key role in two respects. He was serving as chairman of the Academy's division of biology and agriculture at the time, and in this capacity it fell to him to take the lead in choosing members to serve on the Academy review panel. And, since the review had to be done on a rush basis, Norman ended up writing the brief memorandum which served as the Academy's official report to DOD. Thus, from beginning to end, Norman was in control of the project, a fact which casts doubt on the credibility of the whole enterprise. "It was unbelievably improper," says one Academician familiar with the herbicide controversy.

However, Seitz defends Norman as "one of the best division chairmen we've had" and adds: "I've never detected any overt bias on his part." Moreover, Norman contends that the project was handled "with meticulous care." He says that, while his "most important role was in finding people," there were perhaps a dozen other people involved in the selection process. "You might say that I influenced the report in the selection of people, but, believe me, this wasn't a one-man operation," he says. "It would be quite improper of you to suggest that somehow or other I picked the committee. I certainly did not. On the other hand I did, and would again, defend to the president of the Academy my belief that these individuals were competent to do what we were asking them to do."

The committee which Norman and his colleagues appointed was heavily weighted with scientists experienced in the production or use of herbicides and underrepresented with ecologists who would tend to be sensitive to environmental intrusions. The committee included an official of Dow Chemical Company, a major manufacturer of the herbicides used in Vietnam, but it included not a single known critic of the military use of herbicides.* Only one of the six

---

*The panel was chaired by A. S. Crafts, University of California at Davis. Other members were Keith C. Barrons, director, plant science research and development, Dow Chemical Company; Richard Behrens, department of agronomy and plant genetics, University of Minnesota; William S. Benninghoff, department of botany, University of Michigan; William R. Furtick, department of farm crops, Oregon State University; and Warren C.

members, William S. Benninghoff, a botanist, approached the issue from an ecologist's perspective. Norman nevertheless contends that the committee members were competent scientists capable of reaching sound judgments whatever their past professional backgrounds may have been. "You know," he complains, "one of the strange developments of the present time is to say that a person who has been involved deeply in something or other has to be cast aside because he's biased by reason of his past connection. I think this is an enormous mistake—one the government is making rather frequently these days. That a man happened to be employed by Dow Chemical doesn't destroy his credibility as a scientist I would hope." The problem with that argument is that it ignores the need for a balance of viewpoints on the committee, since we are all prisoners of our training to some extent. A committee of Rachel Carsons would presumably come to very different conclusions about herbicide damage than a committee of Dow Chemical scientists. To put herbicide "hawks" on the committee while denying representation to the "doves" would seem from the start to have jeopardized the objectivity of the Academy review.

The MRI Report and the Academy's review of it were carried out in great haste, for DOD was trying to arrange an assessment that would satisfy the AAAS and head off any AAAS drive for further action. MRI completed its part of the project in 3½ months—between August 15, 1967, and December 1, 1967. And the Academy needed only two more months, until January 31, 1968, to complete its review. In striving for early completion, the writing and review functions became somewhat blurred. On November 7, 1967, even before MRI had completed its report, members of the Academy panel met with officials of MRI and of DOD to discuss the project and to receive four completed chapters. Most of the Academy reviewers made comments on the report while it was still in draft stages, and these comments were apparently reflected in the final version of the report. Thus the Academy contributed to the report it was going to review—a fact which undoubtedly made the MRI Report stronger but made the NAS "review" something of a fiction.[17]

Shaw, crops research division, Agricultural Research Service, U.S. Department of Agriculture.

The MRI Report, a massive, 369-page document, was based on examination of more than 1500 pieces of scientific literature supplemented by interviews with more than 140 knowledgeable people. The report focused most of its attention on civilian application of herbicides to non-cropland areas such as roadsides and telephone right-of-ways. It devoted less attention to agricultural use of herbicides or to direct military experience in Vietnam. The report is so large it is difficult to summarize, but its main conclusion was that the greatest ecological consequence of using herbicides in Vietnam or anywhere else is destruction of vegetation, with the result that a region is set back to an earlier stage of development. The report also found that food chains would be altered, with long-term effects on wildlife that might be beneficial or detrimental; that herbicides used in Vietnam would not persist as a phytotoxic level (that is, a level poisonous to plants) in the soil for a long period of time; and that lethal toxicity to humans, domestic animals, or wildlife was highly unlikely. Data were inconclusive with respect to chronic toxicity and many other issues.

The Academy review of the MRI Report was done on a rush basis. The review panel never met as a group after receiving the full MRI Report, a circumstance which made it difficult to compose a report that was truly a committee effort. Instead, the individuals on the committee submitted critiques to Norman, and Norman put together a memorandum which constituted the Academy report on the issue. "There was an enormous time pressure on this, as I recollect," Norman says. "I think they [the committee members] never got together, all of them, in the same time and place, which was not a good way of going about it."

The report which Norman, the former Chemical Corps expert, turned out was headed by a 2¼-page summary memorandum which expressed the consensus of the reviewers as follows:

(1) Midwest Research Institute has done a creditable job of collecting, correctly abstracting and citing much of the relevant published information although, under the circumstances, the report could not be expected to cover in a truly comprehensive way so vast a literature.

(2) Of necessity, the preponderance of the material deals

with herbicides as they are used in vegetation management in a diversity of situations and environments. On this general topic, abundant data are available. However, the scientific literature provides markedly less factual information on the ecological consequences of herbicide use and particularly of repeated or heavy herbicide applications. The Midwest Research Institute Report correctly reflects this disparity.[18]

This summary memorandum was followed by five pages of brief paragraphs excerpted by Norman from the individual reviews submitted to him. The comments ranged from one which thought the MRI Report was being too tough on the military to another which thought that an immediate study should be launched to determine the effects of the defoliation program. In transmitting the Academy review to DOD, President Seitz added a cover letter stating that the MRI Report was "only a first step in investigating further the ecological consequences of intensive use of herbicides." Seitz said the Academy would be "glad to participate in any useful way in the planning and promotion" of further research.[19]

And that was it. That brief package constituted the Academy's entire report. It would be hard to imagine a much blander treatment of a controversial problem. MacArthur later described the Academy review as "a waffling report." A technology assessment study by the Library of Congress described it as "somewhat noncommittal."[20] *The New York Times* said the Academy was "gingerly not taking sides in the controversy."[21] And some critics complained that it was worse than that. "It was a whitewash—a pro forma Good Housekeeping seal of approval," says Arthur H. Westing, chairman of biology at Windham College, Vermont, who has participated in several studies critical of the defoliation program.

The Defense Department, as we have seen, was assuming that the MRI Report and the Academy review of it would prove supportive of DOD's continued use of herbicides in Vietnam, but DOD officials couldn't be sure just what the Academy panel might say. So, just to be on the safe side, DOD beat the Academy to the punch and put out its own version of the MRI findings before the Academy could com-

plete its own review. A 4½-page DOD-prepared summary of
the MRI Report was issued around February 1, 1968, thus
assuring that DOD's version of the report gained circulation
before either the report itself or the Academy's review of it
were made public in mid-February.

Press comment was confused, with some stories saying
there was little likelihood of damage from defoliation in
Vietnam, others saying the effects were unclear, and some
warning that there was indeed danger of long-term damage.
Part of the confusion probably stems from the fact that
some reporters were quoting from the MRI Report itself,
while others relied on the Pentagon-prepared summary.[22]

Comments in the professional press ranged from praise to
condemnation of the MRI Report, but almost all reviewers
agreed that the MRI exercise shed little light on the situation in Vietnam—a crucial point that the Academy had
largely ducked on the ground that it was not supposed to
consider that problem. Fred H. Tschirley, an herbicide specialist with the U.S. Department of Agriculture who has
conducted defoliation research for the military, found the
MRI Report "well done" but "disappointing because its
direct applicability to Vietnam is so tenuous."[23] Frank E.
Egler, an ecologist, complained that "Despite the Pentagon's contention that there are no long-term ill effects for
Vietnam, on its population or its environment, we simply do
not know. It is the differences, rather than the similarities,
of herbicide use in America and in Vietnam which give rise
for concern."[24] Howard T. Odum, another ecologist, stated
that "since data on Vietnam ecosystems are not in this
volume, verdicts on those herbicide actions must be held for
the future."[25] And Sheldon Novick, editor of *Scientist and
Citizen* (now *Environment*), complained that the MRI
Report "actually deals very little with the question of herbicide spraying in Vietnam" but might "contribute to the
general confusion on this topic" by generating "an impression that there is direct evidence available concerning the
Vietnam spraying, where in fact there is none."[26]

Thus the first round of the Academy's participation in the
defoliation controversy—the review of the MRI Report—
had ended, and the Academy had said little that would shed
light on whether the program was causing serious ecological
or health problems. Why was the Academy's performance so

lackluster? One explanation probably lies in the fact that the man who put the Academy's memorandum together was a former developer of herbicides for the Army Chemical Corps, while the reviewers who submitted comments to him were primarily men who used or studied herbicides as a useful agricultural tool and thus would tend to view them in a positive light. Another explanation is that the very restricted role accepted by the Academy made it virtually impossible to say anything significant. "Where the confusion has arisen is that the Academy has been pictured as doing a poor job on something much larger, which in fact we were not asked to do," says A. G. Norman. "It was all a lot of nonsense," snorts Academician George Wald, a Nobel Laureate. "The issue was the long-term effects of herbicides in Vietnam. But the MRI people hardly even left Kansas City. It was all done in the library. All the Academy review could do was say, 'Yes, it's a creditable library job.'"

The MRI Report and the Academy review of it failed to satisfy the AAAS. One of the first actions the AAAS took after getting the two documents was to send them out for comments by its own reviewers. The judgments expressed by these men seemed considerably more harsh than those expressed by the Academy's reviewers. Jean Mayer, the Harvard nutritionist, found the MRI Report "grossly inadequate as an answer to the questions raised by Professor Pfeiffer and myself. It does not cover any aspect of the direct action of the compounds on human beings. . . . It does not cover any of the public health and the human ecology aspects of our crop destruction program. . . . It does not cover any of the social and psychological aspects of this operation . . . little or no attention is paid to the effects on health of disturbing the ecological balance." E. W. Pfeiffer regretted that "so few ecologists of stature" were involved in preparation of the MRI Report and complained that neither MRI nor the Academy had dealt effectively with long-term effects. Ecologist Frank Egler particularly attacked the Academy's review memorandum as "inadequate, in that it evades all the main issues of the problem." George Sprugel, Jr., chief of the Illinois Natural History Survey, concluded that, while MRI had done a "rather thorough" literature review, the whole exercise indicated "inadequate data upon which to develop a reliable assessment of the long-range ecological effects of herbicide use." James P. Lodge, a chemist at the National Center for Atmospheric Research in Boulder, Colorado, found the MRI Report a "very impressive document" but came away from it feeling that there are "frightening" possibilities of "an accident involving the broadcast dissemination of a pesticide."[27]

These individual reviews were never published, but the AAAS board itself, after reviewing the comments of these consultants and after conducting an extended review of its own, issued a policy statement on July 19, 1968. The board almost unanimously agreed to a basic statement expressing its "conviction that many questions concerning the long-range ecological influences of chemical herbicides remain unanswered." In order to resolve all fears and doubts, it urged that a field study be undertaken under United Nations auspices and that "the maximum possible amount of relevant data be released from military security, so that the scientists conducting the study may know the areas affected, the agents used, the dates applied, and the dosages employed."[28]

Unfortunately, neither the United Nations, the Department of State, nor the Department of Defense showed much interest in having a field study undertaken. So the AAAS, slowly and agonizingly, set about organizing its own investigation. It was not until late 1969, however, that the board was finally able to announce that it had named Matthew S. Meselson, a Harvard biologist, a distinguished member of the honorary NAS, and a longtime critic of chemical and biological warfare, to head the effort.[29] There was no thought that Meselson would direct a full-scale field investigation; rather, he was to organize a group that would prepare a detailed plan for such an investigation and would make a "preliminary study" of its own in Vietnam.[30] Operating on a slim budget of about $70,000 provided by the AAAS (later supplemented by grants of about $10,000 from the Ford Foundation and $27,000 from the National Institute of Environmental Health Sciences), Meselson and three colleagues reviewed the pertinent scientific literature, consulted with dozens of experts, and made a six-week inspection tour to Vietnam in August and September of 1970.

The AAAS team received considerable logistics support

from the U.S. government and was given access to top civilian and military officials. But the Defense Department refused to declassify data on the precise time and location of defoliation missions, including the identity and amount of herbicide sprayed. Meselson considered his failure to obtain such information "a major disability" because lack of such information made it difficult to compare effects in areas that had been sprayed with effects in areas that had not.[31] The Defense Department also denied the AAAS team access to data from an Army medical group's survey of stillbirths and abnormalities in Vietnam,[32] although that study was later published and thus became available to the team after its return to the United States.

Despite the handicaps, Meselson and his colleagues presented a preliminary report to the December 1970 AAAS annual meeting which added up to a charge that the military use of herbicides had been more destructive than generally realized. Their report asserted that:

—Roughly half of South Vietnam's mangrove forests had been sprayed, with the result that: "Essentially all vegetation is killed. Preliminary aerial and ground inspection ... showed little or no recolonization by mangrove tree species after three or more years."

—Roughly one-fifth of South Vietnam's merchantable hardwood forests had been sprayed. Aerial inspection over a wide area "showed more than half of the forest to be very severely damaged. Over large areas, most of the trees appeared dead and bamboo had spread over the ground."

—Some 2,000 square kilometers of land had been sprayed to destroy food crops, mostly in the food-scarce Central Highlands, inhabited by Montagnard tribes. "Nearly all of the food destroyed would actually have been consumed by [indigenous civilian] populations" rather than by enemy soldiers.

—There was no definite evidence of adverse health effects, but further study is needed to determine the reason for a high rate of stillbirths in one heavily sprayed province and for an increase in two particular kinds of birth defects which were reported at a large Saigon hospital and which were coincident with large-scale spraying.[33]

The report had immediate political impact. On December 26, 1970, the very day the AAAS convention opened, the

White House announced that authorities in Saigon were "initiating a program for an orderly, yet rapid, phaseout of the herbicide operations."[34] Meselson had previously briefed the State Department on his findings and had given the White House an advance summary of the report he was about to present to the AAAS.

Well before Meselson came out with his preliminary report, which included numerous suggestions for further research, supporters of the Pentagon on Capitol Hill were taking steps to insure that any more ambitious study of defoliation would lie in the more friendly hands of the Academy. The Senate Armed Services Committee, long known for its close relationships with the military, attached a provision to the military authorization bill for procurement for fiscal 1971 which required the Secretary of Defense to ask the Academy to conduct a "comprehensive study and investigation" into "the ecological and physiological effects of the defoliation program carried out by the Department of Defense in South Vietnam."[35] The provision was inserted into the bill by Senator Thomas J. McIntyre (D-N.H.), chairman of the Subcommittee on Research and Development of the Armed Services Committee, who had received informal assurances that the Academy would be receptive to the task.

The chief motivation for this provision seems to have been fear that, unless such a study were ordered, some more drastic action might be taken against the herbicide program. This became apparent in August 1970 during debate in the Senate over the military procurement bill. Members of the Armed Services Committee successfully used the fact that the Academy would conduct a study as a weapon to beat back an amendment by Senators Gaylord Nelson (D-Wisc.) and Charles Goodell (R-N.Y.) to ban the herbicide program entirely. The floor fight against this amendment was led by McIntyre, a late-blooming dove on Vietnam, who had been instrumental in engineering previous compromise cutbacks in the chemical and biological warfare program. McIntyre said he favored the use of herbicides because they had "saved the lives of Americans in Vietnam." But he acknowledged that he and the Armed Services Committee had doubts about the possible side effects of herbicide spraying. Consequently, he said, the committee proposed an "authoritative" study by an "unimpeachable group," which turned out to be the Academy.[36] Senator John Stennis, chairman of the Armed Services Committee, backed up McIntyre by arguing that the bill his committee had recommended "fully recognizes the problem of the use of herbicides. It calls for a study by the National Academy of Sciences. . . ."[37] This pitch was apparently effective. The Nelson–Goodell amendment was rejected 67 to 22. And a subsequent attempt by Nelson and Goodell to ban the anti-crop program was defeated 48 to 33. The legislation providing for the Academy study was ultimately approved on October 7, 1970.

Though the legislation was intended to head off something worse, it still left the Department of Defense in the position of being the target of an outside study that could embarrass it. Ordinarily, one might expect that the Pentagon would be apprehensive, but Pentagon officials actually welcomed the Academy study and figured they could use it to good advantage—to blunt the AAAS critique and to plan for future herbicidal warfare. At a briefing in early 1971, a top Pentagon scientist told Senate staffers that there were "three good reasons" to go ahead with the Academy study even though the herbicide program was declining in Vietnam as the United States withdrew its forces. The reasons, as described in a privileged memo summarizing the briefing, were:

"1. We would have a solid base of facts in case herbicides were considered for use in future conflicts.

"2. It should be known exactly what damage has been done so we'll know how to repair it.

"3. Something should be done in response to charges by AAAS—the American public deserves to know the facts."

It's only fair to note that the Academy itself was not developing plans for herbicidal warfare; it was simply supposed to study the ecological and physiological effects of the spray program. But the Pentagon simultaneously launched its own study of the military benefits of herbicides in order that military planners, by combining their own findings as to military effectiveness with the Academy's findings on side effects, could make an overall judgment concerning the wisdom of resorting to herbicidal warfare. The analysis of military effectiveness, which was carried out by the Army Corps of Engineers in late 1971, gave a qualified endorsement to use of herbicides and recommended that they be included in contingency plans for possible conflicts in such areas as Western Europe, Cuba, Venezuela, Ethiopia, and Korea.[38]

Meanwhile, the Academy, in organizing its field study of side effects, managed to avoid the apparent conflicts of interest which had marred the earlier Academy review of the MRI Report. The chairman of the seventeen-member study committee was Anton Lang, director of the Atomic Energy Commission's Plant Research Laboratory at Michigan State University, while the staff director was Philip Ross, a former National Institutes of Health official. Neither had previously been involved in the military herbicide program. The Academy deliberately excluded from the committee any scientists—such as Meselson and Pfeiffer—who had previously criticized the herbicide program as well as any scientists from the military or the chemical industry. The committee did include at least two civilian scientists who had previously helped in the military use of herbicides—Fred H. Tschirley, a tropical plant expert who had directed herbicide studies under contract with the Pentagon's Advanced Research Projects Agency, and Geoffrey E. Blackman, a British agricultural expert who had advised the British military on defoliation in Malaysia. But the critics of herbicides were also given significant input when Meselson was appointed to the internal group that reviewed the Lang committee's report before it could be released. Thus, while the study group itself appeared slightly favorable toward the military use of herbicides, the review panel—largely because of Meselson's presence—appeared to lean slightly the other way.

The Lang committee, assisted by some thirty consultants, unquestionably conducted the most thorough study yet of the defoliation campaign. It operated on a budget of about $1.4 million and spent some 1500 man-days on the scene in Vietnam, thus finally providing the field study long sought by concerned scientists. But its efforts were hampered by continuing military action in Vietnam, which made it dangerous to visit large areas of the country. Moreover, American military officials were occasionally uncooperative. Although the committee got virtually all the data it asked for, in contrast to the earlier experience of the AAAS group, military authorities were so slow in delivering some of the material that the committee was unable to carry out the "high priority" task of relating herbicide operations to population distribution, a correlation considered crucial for calculating health effects.[39]

After more than two years of work, and considerable internal bickering, the committee issued its findings in early 1974 in a four-hundred-page summary volume backed up by nineteen working papers. The report is said to have been toughened up considerably during the review process within the Academy. Early drafts were reportedly lenient in assessing damage due to herbicides, whereas the final report finds some serious damage, although not as much as some critics had alleged.[40]

The heaviest damage was found in the mangrove forests along the coast. The committee estimated that about 36 percent of the mangroves had been destroyed or very severely damaged (the AAAS had estimated 50 percent) and that it might take more than a century for much of the devastated area to be reforested.[41]

The committee also found "extensive and serious" damage to the inland forests. But its estimate of the loss of merchantable timber was much lower than previous estimates. The committee was unable to inspect these forests on the ground, so its estimate was based almost entirely upon interpretation of aerial photography, a difficult and uncertain art at best. After much internal dispute, the committee concluded that about 1.25 million cubic meters of merchantable timber had been killed by herbicides, out of a total stock of 8.5 million cubic meters of such timber. Even that estimate was deemed too low by two committee members, an independent expert, and Meselson.[42]

On the sensitive issue of human health effects, the committee found little conclusive evidence of significant harm. An anthropologist who interviewed Montagnards living in the highlands of South Vietnam and in refugee camps heard consistent reports that children, and sometimes adults as well, became ill or died after their area was sprayed. The committee judged these reports "so striking it is difficult to dismiss them." But, since the committee had not started its work until the spray campaign was over, it was unable to conduct direct medical observations of the afflicted populations at the time of spraying to confirm or refute these reports. President Handler dismissed the reports as "secondhand . . . tales," a description which enraged the anthropologist who had ventured into the Montagnard villages to conduct his interviews. On other health issues, the committee concluded that available data "does not support the suggestion that herbicide spraying may have engendered birth defects"; that Meselson's reported discovery of an extraordinarily toxic compound, dioxin, in fish and shellfish from Vietnamese waters raised "serious, legitimate concerns for the public health" which should be further investigated; and that defoliation may have led indirectly to the introduction of malaria-bearing mosquitoes, a possibility that required further investigation. But President Handler, in a cover letter that some felt downplayed the hazards, concluded: "On balance, the untoward effects of the herbicide program on the health of the South Vietnamese people appear to have been smaller than one might have feared."[43]

Reaction to the report was mixed. The initial press accounts tended to stress the severity of the damage done by herbicides—perhaps because reporters were getting much of their information through leaks from herbicide critics even before the Academy report was made public.[44] This was a marked reversal of the situation in 1968 when the Pentagon controlled the early publicity on the MRI Report. Both sides in the herbicide debate used the Lang report to bolster their cases. Senator Nelson said the report confirmed his belief that the nation must renounce herbicide warfare.[45] But the Pentagon said a close reading of the report revealed that "some damage has resulted from the military use of herbicides in Vietnam, however, most of the allegations of massive, permanent ecological and physiological damage are unfounded."[46] Academician Kenneth V. Thimann, a supporter of the defoliation campaign, contended that the Lang committee "completely failed to find evidence to support the claims of the earlier . . . AAAS commission."[47] But Meselson believes the two groups reached "remarkably similar conclusions."

From the viewpoint of this study, perhaps the chief revelation to emerge from the defoliation controversy has less to
do with Vietnam than with the institutions of science. The
lesson seems to be that the Academy can't always be
depended upon to tackle a sensitive issue and provide an
objective answer. Indeed, the performance of the Academy
stands in sorry contrast to the performance of the AAAS on
this issue. This is not to say that the AAAS project was
perfectly executed. The AAAS study was marred by two
high-level resignations from one of its early planning committees,[48] by charges that one member of Meselson's committee was particularly biased,[49] and by seemingly endless
delays (the time span from Pfeiffer's original resolution in
1966 until Meselson completes his final report will be more
than eight years). But in a broad sense the AAAS has performed the role of responsible citizen. It gave institutional
voice to the genuine concerns over defoliation in the scientific community; it exerted pressure that forced DOD to
commission the MRI Report and the Academy review of it; it
continued to press for a field study; and, when no one else
would undertake one, the AAAS reluctantly decided to go
ahead and sponsor an independent study itself even though
such a role was foreign to it. Moreover, the study obviously
had impact in spurring a cutback on crop destruction and
defoliation.

In contrast, the Academy expressed no interest in the
herbicide issue until it allowed DOD to use its prestige to
bolster the MRI project. On that occasion, it put the review
in the hands of a former herbicide program official, set up a
committee that seemed biased toward the users of herbicides, and issued a wishy-washy report that failed to grapple
with the issues AAAS had raised. Then the Academy lapsed
into silence until, once again, it proved useful to DOD and its
allies on Capitol Hill to bring the Academy back into play in
another effort to blunt criticism of the herbicide program.

This whole episode has shown that the Academy and the
AAAS are listening to different drummers—the Academy
tends to respond to the needs of program managers in government, while the AAAS tends to give voice to the concerns
of the scientific community. There is no monopoly on wisdom
in these matters, so the AAAS voice is a welcome addition to
the chorus of Washington advice-givers. We hope other

professional societies will also consider voicing their opinions on controversial issues, for the fairest overall assessment is apt to come from differing sources of technical advice. It seems clear that, had the herbicide problem been left solely to the Academy, the pertinent questions would not even have been asked.

CHAPTER EIGHT

The Food Protection Committee:

Protection for Whom?

"It's a joke that these people are being passed off as unbiased experts. They go overboard in their effort to whitewash."

—Jacqueline Verrett, Food and Drug Administration scientist specializing in the study of birth defects, in an interview, 1972.

"I am worried personally about the way the committees of the National Research Council are set up. We all know that you can always set up a committee of scientists to reflect a certain trend."

—Umberto Saffiotti, associate scientific director for carcinogenesis, National Cancer Institute, quoted in *Science*, August 18, 1972.

A panel of the Academy's Food Protection Committee was meeting in February 1972 to evaluate the safety of a red food coloring when an incident occurred which sheds light on much of the committee's recent work. Jacqueline Verrett, a scientist at the Food and Drug Administration (FDA) who is one of the world's leading authorities on the use of the chick embryo to detect chemical hazards, was called in to present the results of her studies of the food dye known as FD&C Red No. 2. She explained that, while her work was not yet completed, preliminary evidence indicated that Red 2 was toxic to the chick embryo at both low and high concentrations, with a lesser effect at intermediate doses. When Dr. Verrett had finished her presentation, Julius M. Coon, the chairman of the Academy panel, commented: "Well, we all appreciate Dr. Verrett's coming over here and entertaining us this afternoon."

What had Coon meant when he thanked Verrett for "entertaining" the committee? According to Coon, his comment was simply a maladroit way of expressing interest in her work. "Sometimes I use words a little recklessly," he told us. But from Verrett's perspective the comment was much less innocent. She regards it as a revealing slip of the tongue which confirms her belief, based on years of dealing with the Academy, that the Food Protection Committee is scornful of scientists who raise questions about the safety of chemicals in the food supply. "I knew they considered it a joke but you'd think they'd go out of their way not to make it obvious," she said.

Verrett's interpretation of the particular incident is open to debate, for only Coon knows for certain what he really meant. But Verrett's broader criticism of the committee—a subordinate unit of the Academy's Food and Nutrition Board which plays a key role in advising FDA on the safety of our food supply—is echoed by a number of scientists both inside and outside the Academy. The general thrust of most criticism is that the Food Protection Committee has been lenient in assessing the hazards associated with various food chemicals. A number of reports issued by the committee between 1968 and 1972 reflect a consistent bias—they hesitate to recommend stern action that might restrict the use of chemical additives, and they downplay the likelihood of subtle, long-term hazards, such as cancer, genetic damage, and birth defects.

There appear to be two main reasons for the committee's leniency. One may be that the committee has long had very close ties with the food and chemical industries. The committee was originally established in 1950 along guidelines suggested by an organizing committee composed largely of industrial representatives, and it has consistently received financial support from the food, packaging, and chemical industries.[1] In fiscal year 1972, the committee received $78,-000 in support from industry—about 40 percent of its total budget for that year. The industry funds support part of the cost of the permanent staff and pay for studies that the committee launches on its own, while the government pays for studies that it specifically requests. The committee also has an industry liaison study panel which helps it get industrial information that might otherwise be difficult to obtain.

At the same time, the panel gives industry a voice in committee affairs. It often suggests ideas for possible studies; it sometimes pressures the Academy to undertake studies against its better judgment; it sometimes exhorts the committee to defend industrial interests; it proposes candidates to serve on committee panels; and individual members of the industry liaison group are sometimes asked to review committee reports (other than those done specifically for the government) before they are released. Academy officials stress that industry does not "control" the committee in the sense that it can dictate who will be named to panels or how a report will be worded. Indeed, they insist that industry is sometimes unhappy about projects that are undertaken—one such case allegedly being a report on "toxicants occurring naturally in foods," a subject which industry officials felt might unduly alarm the public. But the industry viewpoint unquestionably gets an extremely sympathetic hearing at the committee—probably more so than in most Academy committees. Marvin Legator, former head of FDA's genetics toxicology branch, told us: "They're biased as heck for industry."

The second reason for the committee's leniency is that it has long been dominated by scientists who tend to be more concerned about immediate, acute effects than about long-term, chronic effects. The committee's panels have seldom, if ever, included anyone primarily expert in mutagenesis (the production of genetic defects) or teratogenesis (the production of birth defects), and the members expert in carcinogenesis (the production of cancer) have been a small minority. Yet in many cases the most alarming evidence of hazards associated with a particular chemical has involved precisely these long-term effects. Thus, on a matter of great importance to every American—the safety of the food supply—the committee has often issued reports that ignore the insights and evidence of some of the most relevant scientific disciplines.

Some of the most influential scientists on the committee in recent years have publicly opposed critics of environmental degradation and other long-term insults to human life. William J. Darby, a Vanderbilt University biochemist who headed the committee from 1953 to 1971, was one of the most intemperate critics of Rachel Carson. Similarly, Julius M. Coon, a Jefferson Medical College pharmacologist who headed many of the Academy studies that have been criticized, has openly derided contentions that food additives may cause cancer, birth deformities, or genetic defects. "There is not a shred of evidence or even a basis of reasonable suspicion that any such damaging effects have ever been caused by the additives or pesticides in food consumed in North America," he wrote in 1970. "Certainly some defects have been observed in test animals after they have been fed exceedingly large amounts of some additives. But it is a long, frequently too long, step from the observation of the effects of such provocative and bizarre experiments to those of man's daily diet."[2] And Leon Golberg, an Albany Medical College toxicologist who is another committee stalwart, has suggested that scientists who claim to find mutagenic, carcinogenic, and teratogenic effects go out of their way to devise absurd tests that will undermine the food industry. "Carried to their logical conclusion, these considerations of carcinogenesis, teratogenesis and mutagenesis strike at the very roots of technological advance in food production."[3]

The views of such scientists have had a persisting impact on the Food Protection Committee because a relative handful of scientists has dominated the committee's studies year after year. "It's always the same tired old faces," comments one FDA official who has had long experience dealing with the Academy. "They have not been keeping up with the literature and they have not had any new thoughts. They're just not with it." That situation is changing as a result of Handler's new policy of rotating people off committees after a specified term of service. Darby, the most influential voice on the committee for years, was forced off the main committee in 1971 because he had served such a long time as chairman. Similarly, Coon was dropped from the main committee in 1972, though he remained active in subcommittee work. Nevertheless, the committee remained unbalanced. Its thirteen members in 1972–1973 included no one whose primary interest was genetics or teratology and only one man whose primary expertise was cancer. In a September 1972 interview, Handler acknowledged that the committee was lacking expertise with respect to chronic effects; he said he would broaden the membership to include new viewpoints.

The committee's tendency to pass over long-term effects was apparent in a 1969 report entitled *Guidelines for Estimating Toxicologically Insignificant Levels of Chemicals in Foods*. That report attracted little attention at first, but it eventually became the target of sharp criticism.

The report was produced by a special task force composed primarily of industrial scientists and traditional toxicologists. Of the nine members of the task force, five were directly employed by food or chemical companies and a sixth headed a commercial laboratory which serves the food and chemical industries. These are the very industries which have been responsible for introducing chemicals into the food supply and whose profits would be affected by a get-tough attitude toward chemicals or by a requirement that all chemicals be extensively tested. The other members of the committee were Coon and Darby, whose antipathy toward antichemical crusaders has been noted here, and the late Henry F. Smyth, who had had long experience in industrial hygiene. There was not a single scientist on the task force whose primary expertise was in carcinogenesis, mutagenesis, or teratogenesis. Nor was there anyone who could be identified as a champion of consumer, as opposed to corporate, interests.*

The report turned out by this task force contended that "for every chemical" there is a finite level "at or below which it can be present in food without prejudicing safety."[4] This seemed to be an endorsement of the so-called threshold theory which holds that a chemical can be deemed dangerous above a certain threshold level but safe below that level. The theory is not accepted by the vast majority of cancer experts and geneticists, who contend that, given the current state of knowledge, it is impossible to set any safe threshold for chemicals that cause carcinogenic or mutagenic effects.

The main purpose of the report was to set forth guidelines to help determine which chemicals are so innocuous that they can be allowed in the food supply in "toxicologically insignificant" amounts even though they have not been tested for safety in the laboratory.†

---

*The task force was headed by Henry F. Smyth, Jr., a toxicologist at the Mellon Institute. The other members included Julius M. Coon, chairman of the department of pharmacology at Jefferson Medical College; J. P. Frawley, chief toxicologist at Hercules, Incorporated, a major chemical company; Richard L. Hall, vice-president of McCormick & Company, Incorporated, a major food company; Bernard L. Oser, president of Food and Drug Research Laboratories, Incorporated, a commercial lab serving primarily industrial clients; Arthur T. Schramm, Food Materials Corporation; John A. Zapp, director of the DuPont Company's Haskell Laboratory; William J. Darby, head of the biochemistry department at Vanderbilt University and a science adviser to Corn Products Company; and Richard Henderson, of Olin Mathiesen Chemical Corporation.

---

†The standard technique for determining the "safe level" of a chemical is to conduct laboratory tests and then apply a safety factor to the results. Experiments are typically performed to find the highest dietary level of a chemical that has no adverse effect on test animals, and this level is then divided by 100 to obtain an assumed "safe level" for the chemical in the human diet. The 1:100 safety ratio is applied because man may be more sensitive than the test animals and because, in any large population, some individuals may be unusually susceptible to harm from the chemical. This traditional safety ratio is a rule of thumb rather than a mandatory requirement; a higher or lower ratio is sometimes used if experienced judgment suggests it would be more appropriate.

Unfortunately, this traditional approach to determining safe levels has a practical drawback: there are tens of thousands of natural and synthetic chemicals found in the food supply and it would be extraordinarily difficult,

The report proposed several criteria for making such judgments. One was based on the length of time a chemical had been in use. The committee concluded that if a chemical had been in commercial production *for five years or more* without evidence of toxicological hazard, and, if it did not fall into certain particularly dangerous categories, then "It is consistent with sound toxicological judgment to conclude that a level of 0.1 ppm of the chemical in the diet of man is toxicologically insignificant."[6]* Another criterion was based on comparison with similar chemicals. The committee concluded that a level of toxicological insignificance could be established for a chemical simply by extrapolating from the results of studies performed on "at least two analogous substances" which exhibit similar "acute or subacute toxicity."[7]

These guidelines are potentially very influential, for the FDA's own regulations require that "The Commissioner will be guided by the principles and procedures for establishing the safety of food additives stated in current publications of the National Academy of Sciences–National Research Council."[8] The commissioner does not have to adhere blindly to the Academy's recommendations—in this instance the FDA has not yet accepted the numbers proposed by the Academy. But the Food Protection Committee is clearly in a strong position to influence the agency's thinking.

Although little noticed at first, the committee's guidelines eventually provoked dismay in some segments of the scientific community. Samuel S. Epstein, professor of environmental health and human ecology at Case Western Reserve University, an expert on toxicology with particular reference to mutagenesis and carcinogenesis, has said that the report is "one of the most pernicious documents ever to

---

if not impossible, to test them all on a crash basis. Thus the Academy report sought to establish "a reasonable system of priorities" for determining which chemicals should be tested and which could be assumed safe, or at least assigned a low priority for testing, on the basis of experienced scientific judgment.[5]

---

*The "toxicologically insignificant" level and the "safe" level are not synonymous; the "toxicologically insignificant" level is generally set at a fraction of the assumed "safe" level.

emerge from the NAS–NRC. It never once mentions carcinogenicity, mutagenicity or teratogenicity. It's a terrible piece of industrial apology." Epstein also criticized a followup 1970 report, *Evaluating the Safety of Food Chemicals*.

An even more devastating critique was issued by a committee of cancer experts assembled by the Department of Health, Education, and Welfare (HEW) to consider the problem of environmental carcinogens. That committee was chaired by Umberto Saffiotti, associate scientific director for carcinogenesis of the National Cancer Institute; it included seven other government and academic scientists. The main thrust of the Saffiotti report contradicted almost everything the Academy committee had seemed to say. The report concluded, "No level of exposure to a chemical carcinogen should be considered toxicologically insignificant for man.... The principle of a zero tolerance for carcinogenic exposures should be retained.... Exceptions should be made only after the most extraordinary justification." It also estimated that "The majority of human cancers are potentially preventable" (by protecting people from exposure to carcinogens) and recommended that some 20,000 compounds be tested for carcinogenic effects at an estimated cost of $1 billion.[9]

The Saffiotti committee specifically condemned the Academy report in language that was remarkable for its harshness. It called various parts of the report "scientifically unacceptable," "of dubious merit," and of "absolutely no validity in the field of carcinogenesis." The Saffiotti group particularly criticized some of the criteria that the Academy had proposed for determining toxicological insignificance:

> To assume (a) that a 5-year period of use has any meaning for the evaluation of chronic toxicity in man, (b) that any chemical may be considered safe simply because two analogous substances are "safe," and (c) that acute or subacute toxicity are reliable guidelines for evaluating long-term toxicity is to display a lack of understanding and appreciation of factors involved in chronic toxicity, particularly of the irreversible and delayed toxic effects which occur in carcinogenesis.... [T]he lack of consideration of irreversible long-term toxic effects (which would not be ruled out by the suggested criteria) makes the suggested approach practically inapplicable and potentially dangerous.[10]

The indictment was startling in its tone. The Academy is not accustomed to having its reports described in such terms by an eminent group of scientists. The cancer group's critique, which was incorporated into an official government report after due deliberation, appears to be the most devastating public attack on any Academy report in recent years.

On July 15, 1970, Surgeon General Jesse Steinfeld offered the Academy an opportunity to respond to the Saffiotti report, and the Food Protection Committee subsequently prepared a sharp rebuttal that sought to justify its previously articulated line of reasoning. When this rebuttal reached the Academy's reviewers, it was stopped cold. "We felt the position of our Food Protection Committee was far more questionable than the position of the Saffiotti group," George B. Kistiakowsky, chairman of the Report Review Committee, told the study team. "We recommended against letting the Food Protection Committee send its rebuttal out because it was based on unproven assumptions."

Faced with conflicting views from the Food Protection Committee and the reviewing panel, President Handler ended up blocking the committee's rebuttal and instead prepared a new letter to the Surgeon General which sought to present a compromise position. The letter stressed that "no categorical statements are rationally acceptable" in the complex field of chemical carcinogenesis. It partially supported the Report Review Committee by acknowledging that "We do not as yet have the capacity adequately to assess the hazard to man from potential chemical carcinogens. . . . We do not, for example, know the dose–response curve of any chemical carcinogen [the threshold issue] let alone know whether a given dose–response pattern is characteristic of a whole range of materials [the analogous substances argument]." But with a bow to the Food Protection Committee it added: "Sometimes a near-hysteria on the part of the general public, and at least a portion of the professional and political community, can easily lead us to overreact to situations of possible human hazard," with the result that needed substances may be removed from the market.[11]

Handler urged a "concerted effort to steer a course between the two extremes," weighing each case on its merits until an effective research program could provide more definitive information upon which to rely. He attacked the Saffiotti report's suggestion that some 20,000 chemicals be screened at an estimated cost of $1 billion or more on the grounds that such full-scale testing was "totally impractical." And he disputed Saffiotti's view that all carcinogens should be held to a zero tolerance level.

But Handler's only response to the Saffiotti group's harsh criticisms of the "toxicological insignificance" report was to say that the guidelines were intended as a basis for establishing priorities as to what chemicals most need testing; they were not intended to argue that any chemicals should be perpetually exempt from testing.

In an effort to smooth over the conflict, the Academy hosted a "peace meeting" between members of the Saffiotti group and of the Food Protection Committee, on January 21, 1972.[12] Handler later told Congress that "The two groups were not in conflict" and that the disagreement stemmed largely from "semantics."[13] But interviews with Darby and Saffiotti months after the peace meeting indicated that a fundamental difference of opinion remained. Saffiotti still contended that the criteria set forth in the Academy's report would not be adequate to screen out chemicals that might prove carcinogenic. Darby contended that the guidelines, while not foolproof, could indeed be used to determine the likelihood that a chemical would be toxic in any respect, including the likelihood that it would produce cancer or genetic defects. "You can certainly look at [a chemical] and say there's no reason at all to believe it's carcinogenic and so it's low on the totem pole for carcinogenic testing," he said. "As a matter of fact, this is what I think they do."

As for charges that the Academy panel failed to include a single scientist whose primary expertise lay in the areas of carcinogenesis, mutagenesis, or teratogenesis, Darby contended that the panel had ample access to expertise in these areas. He said that all first-rate toxicologists are familiar with the problems of testing for carcinogenesis and teratogenesis and that the panel drew on a variety of insights by reading and by consulting with experts in the relevant disciplines. But the furor aroused by the report seems evidence that the committee did not, in fact, pay sufficient attention to long-range health problems. Instead, it produced a set of guidelines which, in the opinion of the dissenting experts, could allow dangerous chemicals to remain in the food supply.

The Academy's evaluation of the artificial sweeteners known as cyclamates in reports issued during the late 1960s and 1970s provides further evidence that the food protection panels have been dominated by a restricted toxicological viewpoint which downgrades the likelihood of subtle, long-term hazards. Moreover, the Academy's earliest reports on the sweeteners, issued between 1954 and 1962, reveal that the Academy was unable or unwilling to goad the FDA into significantly controlling the use of cyclamates.

The artificial sweeteners have been widely consumed by both dieters and the general population. Saccharin was the first such sweetener to gain widespread acceptance. It was introduced in the late nineteenth century and was used as a sugar substitute by diabetics, obese individuals, and others who wished to restrict their intake of carbohydrate or calories. Saccharin remained the chief artificial sweetener on the market until the 1950s, but it had a serious drawback that hindered its acceptance: many consumers complained of a bitter aftertaste. In 1950, Abbott Laboratories, a drug manufacturer, introduced a competitive sweetener composed of cyclamate. The cyclamate compounds have much less sweetening power than saccharin but they also lack the bitter aftertaste; they soon became the preferred artificial sweetener. Although the sweeteners were initially limited to special diet foods, they eventually found their way into the general food supply. Under a "grandfather clause" in the Food Additives Amendment of 1958, the FDA placed both cyclamate and saccharin on a list of chemicals "generally recognized as safe" (the so-called GRAS list), a decision which meant that the sweeteners could henceforth be used in any foods in any amounts. This opened the door for a remarkable rise in consumption during the 1960s. The sweeteners were used alone or in combination to replace sugar in a host of products, primarily soft drinks, but also canned fruits, jellies, cookies, salad dressings, gelatin desserts, puddings, frozen desserts, and other items. Both cyclamate and saccharin were also sold in tablet or solution forms that could be added to coffee, cereals or fruits.

Food manufacturers had a strong economic incentive to replace sugar with cyclamates in their products since 64 cents' worth of cyclamates had the sweetening power of about $6 worth of sugar.[14] They also found a receptive market since the American public, thanks to the preachings of nutritionists and physicians, had become increasingly aware of the advantages of leanness as opposed to obesity.

Fueled by a massive advertising campaign, production of cyclamates in this country rose from 5 million pounds in 1963 to 15 million pounds in 1967 and was expected to reach 21 million pounds in 1970—a fourfold increase in just six years.[15] The FDA estimated in 1969 that three-quarters of the nation's population was consuming nonnutritive sweeteners in some amount.[16] Then, in late 1969, cyclamates were found to cause cancer in rats. This forced their removal from the general food supply and left saccharin as the sole sweetener authorized for general use. Cyclamate sweeteners and cyclamate-containing food remained available as nonprescription drugs for almost a year, but on August 14, 1970, the FDA, after receiving a recommendation from a medical advisory group that cyclamates should not even be made available to diabetics on a prescription basis because the risks outweighed the benefits, announced a total ban on the substance.[17]

The early years of the Academy's involvement in the sweetener issue was marked by a series of ineffective warnings. The Academy was first asked to evaluate the safety of the sweeteners in 1954, shortly after cyclamates had come onto the market but well before the massive upsurge in consumption had begun. The Academy's Food and Nutrition Board transmitted a policy statement on artificial sweeteners to the FDA in November 1954, the Food Protection Committee issued a report on the safety of sweeteners in August 1955, and the Food and Nutrition Board made public a slightly revised version of its policy statement in November 1955. The main thrust of these documents was that saccharin and cyclamate were not hazardous when used for special dietary foods but that they should not be used on an unrestricted basis. The reports suggested that saccharin was not apt to be used excessively because it was so intensively sweet and left a bitter aftertaste. But they warned that cyclamate might be used at relatively high levels and that little was known of its effect at those levels. "The priority of public welfare over all other considerations precludes, therefore, the uncontrolled distribution of foodstuffs containing cyclamate,"[18] the 1955 policy statement said.

The only toxic response specifically cited by the Academy was a laxative effect—excessive use of cyclamates (5 or more grams per day) was apt to result in the formation of soft or mushy stools. But the Academy raised questions about the safety of cyclamates for pregnant women, children, or those suffering from kidney dysfunction or lower bowel disease.[19] It also threw doubt on one of the main selling claims of the sweeteners by asserting that "The availability and consumption of artificially sweetened foodstuffs have no direct influence on body weight."[20] (This is because people who think they will lose weight by eating artificially sweetened foods almost always make up for the lost calories by taking extra portions of food and drink.[21]) The Academy also said that, while there was only a "remote" chance that people would consume too much cyclamate in "special purpose" foods, the case was different with general purpose beverages "because individuals may ingest cyclamate in excess of safe limits by normal consumption of cyclamate-sweetened soft drinks."[22]

Despite the Academy's warning, the FDA did little to impede the uncontrolled spread of cyclamate usage. Instead, the FDA's decision to include both cyclamate and saccharin on the GRAS list allowed the sweeteners to invade the general diet of the population. As usage increased, the Academy reviewed their safety once again. A revised policy statement issued in 1962 repeated the warnings of the earlier statement almost verbatim and stressed, additionally, that when the sweeteners were used for other than special dietary reasons their "use should be controlled."[23] But the statement is notable for its lack of urgency. Seven years earlier the Academy had warned that use of the sweeteners should be limited because little was known about long-term exposure at high dosages. In the intervening years few of the questions raised by the Academy in 1955 had been answered and usage of the sweeteners had spread to the general food supply. Yet here was the Academy in 1962 simply reissuing its earlier statement with a few minor revisions. In contrast, other professionals began clamoring for a halt. In 1964, *The Medical Letter*, published by a group of prominent doctors and medical professors, editorialized: "In light of the wide and indiscriminate use of these sweeteners, a reappraisal of the toxicology of artificial sweeteners.... is in order."[24]

The FDA again asked the Academy to evaluate the latest evidence concerning the safety of the sweeteners in November 1967. An ad hoc committee on nonnutritive sweeteners, headed by Julius M. Coon, was assigned the task of reviewing the scientific data. It issued its findings in November 1968 in an "interim report" that proved deeply disappointing to FDA scientists.

The report was timid and equivocal. On the one hand, it concluded that, in numerous studies of cyclamates and saccharin–cyclamate mixtures, "no effects have been observed that can yet justify any serious suspicion of adverse effects at doses or consumption levels of the order of those thought currently to be taken by man under practical circumstances." On the other hand, it warned that "totally unrestricted use of the cyclamates is not warranted at this time" because consumption of 5 grams per day might induce softening of the stools and because there was uncertainty about the toxicological significance of cyclohexylamine, a metabolic derivative of cyclamate. The report concluded that daily intakes of 70 milligrams per kilogram of body weight are "safe."[25] This was less stringent than an earlier recommendation by the Food and Agricultural Organization–World Health Organization (FAO–WHO) that daily intake be limited to 50 mg/kg.[26]

The Academy's report was subjected to unusually careful scrutiny at the FDA because of its regulatory implications and because the FDA had been building up a body of expertise on the sweeteners which enabled agency scientists to evalute the Academy report line-by-line. Two of the agency's top experts—John J. Schrogie, of the Bureau of Medicine, and Herman F. Kraybill, of the Bureau of Science—were asked to direct a review of the Academy report. They had previously prepared a long review of the scientific evidence for FDA's own internal use and were thus thoroughly familiar with the literature on cyclamates and saccharin. Their chief conclusion was that the Academy had done a superficial job. As they saw it:

This report of the Food Protection Committee represents a largely uncritical review of the available material. Conclusions of studies are included without proper regard for the quality of methodology originally used. Studies lacking adequate statistical design are given equal weight with sounder studies; many clinical and epidemiological studies yielding questionable conclusions are given uncritical acceptance. Because such conclusions are included, possibly erroneous results gain added stature and interpretive errors are perpetuated.[27]

The Schrogie–Kraybill critique suggested that the Food Protection Committee had placed undue reliance on usage figures supplied by an industry survey that was "fraught with great hazard of error."[28] It also suggested that the committee had ignored some of the most pertinent evidence, including studies by Jacqueline Verrett, FDA teratologist, which had shown that cyclamates and cyclohexylamine were "teratogenic to the chick embryo."[29] The Academy committee had been sent an "interim report" of Verrett's work,[30] but the committee apparently attached little weight to the data.

The Food Protection panel also ignored the latest FDA findings that moderate amounts of cyclohexylamine caused chromosome breakage in germ cells of the male rat—a warning that the substance was likely to cause genetic damage in humans. The chromosome findings were presented to a scientific seminar in October 1968 by Marvin Legator, head of cell biology research at FDA, about one month before the Academy report came out, and a detailed abstract of the data was provided to the Academy committee before its report was released. But the Academy report made no mention of the findings. Coon told us the data came in too late (after the committee had completed its writing), and was of too preliminary a nature, to permit proper evaluation. The results of the FDA experiments were subsequently published in *Science;*[31] they were judged "presumptive evidence of genetic hazard" by a group of genetics experts, including Joshua Lederberg, a Nobel Laureate, Alexander Hollaender, a member of the NAS, and Epstein.[32]

Thus the Academy came out with a report which suggested that there was little evidence of hazard from the sweeteners at the very time that FDA studies were beginning to raise serious questions about possible cytogenicity and teratogenicity from cyclamates and its metabolite, cyclohexylamine. There was intense debate within the FDA as to what should be done on the basis of the Academy report.[33] But as it turned out, the agency did nothing. On

April 3, 1969, the FDA proposed new labeling requirements for food products sweetened with cyclamates.[34] However, the labeling requirements never went into effect and the status of cyclamates remained unchanged until animal experiments showed they could cause cancer.

Why did the agency fail to act against cyclamates? A congressional committee that investigated the matter found "no satisfactory explanation" unless "lethargy" played a role.[35] The major blame for inaction must fall on FDA, which had the power to move against cyclamate at any time. But the ambiguous, uncritical Academy report did nothing to encourage prompt regulatory steps. FDA Assistant General Counsel William W. Goodrich later testified that "the report was disappointing to us" because of its "waffling" nature.[36]

Why was the Academy report so uncritical? In the opinion of Jack Schubert, University of Pittsburgh radiation chemist and a longtime opponent of cyclamates, the authoring committee was "unqualified" to judge the crucial data. Schubert charged in a public speech that the Academy was "not the type of organization that can pick a good panel"— namely, one that covered "sufficient disciplines" and was free of vested interests.[37] He particularly complained of the lack of a geneticist on the 1968 panel—a complaint which seems well founded. Had geneticists been considering the data it is likely that Legator's data, however preliminary and however late, would have been given more intensive consideration and that a "stop the presses" order would have gone out to permit evaluation of the findings.

The Academy's final assessment of cyclamate came in the fall of 1969, and once again it gave short shrift to possible genetic hazards. The committee on nonnutritive sweeteners was reconvened to consider Verrett's chick embryo studies and Legator's chromosome studies, which had been given scant attention the year before. The committee still lacked expertise in genetics and teratology, the disciplines most relevant to the studies which were to be reviewed, so it called in a recognized authority, James F. Crow, professor of medical genetics at the University of Wisconsin. Crow had impeccable credentials. He was a member of the NAS, the only member of the group to have that badge of prestige. He had also served as chairman of the genetics study section at the National Institutes of Health, president of the Genetics

Society of America, president of the American Society of Human Genetics, and executive officer of the Environmental Mutagen Society.

But before the committee could complete its deliberations on possible genetic and birth defects new evidence became available which indicated that cyclamate was a carcinogen. Abbott Laboratories, the major producer of cyclamates, had been sponsoring long-term feeding studies in animals at a commercial laboratory, the Food and Drug Research Laboratories, Incorporated, in Maspeth, New York. On October 8, 1969, Abbott learned that bladder tumors were showing up in some of the rats being fed high doses of cyclamate and saccharin. Abbott scientists subsequently presented the findings to the National Cancer Institute, the Surgeon General, the FDA, and other government health officials, and a number of cancer experts consulted by the government agreed that tumors had indeed been produced. Nevertheless, the FDA and HEW were still a bit reluctant to remove cyclamates from the market without that ultimate blessing—the imprimatur of the Academy. So, while the Academy committee was busy pondering the data on teratogenicity and mutagenicity, Steinfeld and Ley persuaded the group to consider the latest bladder tumor evidence at a special presentation on October 17.[38]

This referral to the Academy seems to have been done at least partly for public relations purposes. Even Ley acknowledges that the Academy committee "really wasn't" a good group to consult. He notes that the committee was dragged into the situation "completely cold" at the last minute and that it had only one pathologist—even though pathology was probably the key discipline needed to interpret the tumor findings. But he says the FDA wanted to get the committee's reaction because the committee had been studying cyclamates as long as anyone and the agency was faced with the need to take substantial regulatory action on relatively skimpy data.

After reviewing the evidence, the Academy committee prepared a statement the same day expressing its "unanimous opinion" that the material tested—a mixture of cyclamate and saccharin and cyclohexylamine—"was carcinogenic under conditions of the described experiments." The committee said the evidence "strongly" suggested that the culprit was cyclamate or cyclohexylamine but that there was "a slight possibility" that saccharin was involved.[39] In a cover letter sent later to the FDA, Handler stressed that the committee's opinion "rested entirely on the evidence presented. The technical judgment was primarily that of the independent pathologists [those previously consulted by government health officials] who examined the tissue specimens."[40] Although the Academy was clearly uneasy about the way it had been rushed into things, the government invoked the Academy's prestige the next day, October 18, when it announced that cyclamates were being removed from the GRAS list. Steinfeld told a press conference that "after a thorough review" (i.e., one day) the Academy committee "independently confirmed our interpretation of the information and recommended that we remove cyclamates from the list of approved food substances for general use."[41]

Thus the committee found itself in the unaccustomed position of helping to force a product off the market. But later the committee chairman had second thoughts about the matter. "I've always had some misgivings about the whole outcome there," Coon told us. "My personal feeling was—or is now anyway—that we could have said that this was not an appropriate test for carcinogenesis." Coon noted that the study which indicted cyclamate had not been designed as a carcinogenesis study and had not involved cyclamate alone, but a mixture of cyclamate and saccharin. Although the incidence of tumors was statistically significant, Coon said, "it wasn't awfully convincing to some of us."

After the immediate hubbub over the bladder cancer findings had quieted, the Academy committee proceeded to complete its report on the teratogenicity and mutagenicity of cyclamates. It concluded that Verrett's chick embryo studies were "not now persuasive" because tests of teratogenic effects in various mammals had been "overwhelmingly negative." And it concluded that Legator's chromosome tests were suggestive but not conclusive.[42] In a cover letter accompanying the report, Handler stated: "Our committee considers that there is as yet no reliable information concerning the teratogenicity or mutagenicity of cyclamates, and that there is insufficient evidence that cyclamates are a problem in these regards but that there is a body of somewhat suggestive information which indicates that, were cyclamates still on the market, further studies might be called for."[43]

The report and cover letter made no mention of the fact that Crow, the genetics expert called in for consultation, had quite a different opinion. The committee sent Crow a copy of the report before it was released, and Crow wrote back requesting that he not be listed as one of the authors because he disagreed with parts of it. Crow believed that the evidence of chromosome breakage added up to "a fairly strong case that cyclohexylamine, and therefore cyclamates, represent a mutagenic hazard to man." And he expressed alarm over possible teratogenic damage. Crow acknowledged that "there is such a thing as being too careful," but he argued that there was no good reason to take chances with a substance of such marginal usefulness as cyclamates. Further, he suggested that cyclamates should not be used—except where a compensating benefit outweighed the risk—until they were adequately tested for safety. "The inconvenience of not having cyclamates during the period of testing and later finding that they are safe is quite trivial compared to the damage that will already have been done during the testing period if the substance is found to be harmful," he said.[44]

That was quite an indictment from the expert who was specifically called in to bolster the committee's genetics expertise. But no hint of Crow's thinking appears in the Academy's five-page letter report. Crow is listed as a "consultant" to the committee, a designation which conveniently allowed the committee to imply that Crow had contributed to the report without, in fact, requiring that he sign it and agree to it. The episode had little practical significance at the time since cyclamates were ultimately removed from the market. But it illustrates once again how the Food Protection Committee has been dominated by a toxicological viewpoint which gives little weight to subtle long-term effects. "I do share the feeling that this committee and maybe some of the others were not strongly enough represented by people whose interest is in either genetics or long-range teratogenic effects," Crow said in an interview; "I regard this as an omission rather than as a low deed that was deliberately done." Crow suggested that the committee had been put together before it was clear that mutagenesis would be a major point at issue. "I really think that the subject changed faster than the committee did," he said.

The Food Protection Committee's reluctance to ban cyclamates was largely endorsed by Academy President Handler. In a speech to the Biochemical Society in London on December 17, 1969, Handler chided the British for following our "foolish decision to ban the use of cyclamates."[45] And in an April 1970 speech he decried "the precipitate action in banning cyclamates."[46] The "unlikelihood" of cancer in individuals who consume low doses of cyclamate should be weighed, he argued in a 1970 letter, against "the number of lives which are spared or disease which is prevented by the weight-maintaining properties of cyclamate in individuals with a propensity to obesity, or in diabetics."[47]

Handler's views seem sanguine compared to those of a Medical Advisory Group on Cyclamates appointed later by the Department of Health, Education, and Welfare. On the basis of information developed after the Academy's consideration of cyclamates in October 1969, the HEW advisory group concluded that cyclamates were so dangerous that they should probably not be tested in humans. The HEW group also found—contrary to Handler's assertion—that the scientific literature "does not contain acceptable evidence that cyclamate has been demonstrated to be efficacious in the treatment or control of diabetes or obesity."[48]

Handler had defended cyclamates from the supposedly objective and authoritative pulpit provided by the Academy presidency. But he was, in fact, a somewhat biased witness: Handler served from 1964 to 1969 as a member of the board of directors of Squibb-Beechnut, a firm which used cyclamates in its "Sweeta" brand sweetener. Handler told us that he had resigned his directorship and sold his stock in the company when he was elected president of the Academy in 1969. He also said that Squibb had been marketing Sweeta before he joined the board, that his role on the board was to maintain liaison with the research departments at Squibb, and that these departments were not involved with cyclamates. Nevertheless, Handler was a policy maker for a firm that participated in the great upsurge in cyclamates usage. He presumably retains the conditioning of a corporate director who would tend to regard the products of his industry as beneficial and efforts to ban them as unreasonable.

In late 1973, Abbott Laboratories petitioned the FDA to reverse the ban on cyclamates, citing new studies which allegedly showed the sweetener was safe. The FDA was still reviewing the petition in late 1974, but this time, to the relief of cyclamate critics, it did not call on the Academy for an advisory opinion on the matter.

The Academy's evaluation of monosodium glutamate (MSG) in 1970 is chiefly interesting for the light it sheds on how an industry that is under fire for allegedly dangerous practices can successfully use the Academy to ward off criticism. Food manufacturers whose products were threatened with banishment virtually demanded an Academy review; the FDA contracted for such a review; the Academy appointed a committee whose chairman, incredibly enough, promptly started doing research on MSG under sponsorship of the very companies whose products were being questioned; and the committee gave MSG a clean bill of health despite reservations by an expert "maverick" on the committee who felt MSG looked dangerous.

Monosodium glutamate is a popular seasoning agent or flavor enhancer that is widely used in homes and restaurants. It is marketed as a seasoning agent under a variety of trade names and is also used by the food industry as an additive to enhance the flavor of commercially prepared products. MSG is a natural substance—the sodium salt of glutamic acid, an amino acid which is a constituent of all common food proteins. Thus, it has always been part of the normal diet of man and is a normal constituent of human tissue. Among other uses, it is the chief active ingredient of soy sauce, a seasoning agent that has long been used in the Orient. The natural character of MSG, coupled with its long history of use, led the FDA to classify MSG among those substances "generally recognized as safe." As a result, the companies which marketed MSG were not required to conduct tests demonstrating safety and no limits were placed on the amount of MSG that could be incorporated into foods or ingested in the total diet.

In the late 1960s, however, several investigators began raising questions about possible adverse effects from MSG. First there were reports that some adults who had eaten Chinese food suffered painful symptoms—a burning sensation on the chest, neck, and other parts of the body; pressure in the cheeks; chest pain; and occasional severe headaches. The symptoms were described as the "Chinese restaurant syndrome," and the cause was ultimately traced to MSG.[49] No lasting adverse effects were reported, but the surprising discovery of this syndrome after MSG had been used for so many years raised doubts as to how safe the substance actually is. Then reports began surfacing that the MSG used in baby food products might endanger the health of infants. John Olney, associate professor of psychiatry at Washington University in St. Louis, reported that MSG caused brain and eye damage in young mice and suggested that baby foods containing MSG might cause similar damage in human infants.[50] Jean Mayer, professor of nutrition at Harvard University and special consultant in nutrition to the President, told a press audience that, if it were up to him, he'd "take the damn stuff out" of baby food.[51] And consumer champion Ralph Nader criticized the continued use of MSG without adequate study of its effects.[52] The burden of most complaints was that MSG's safety was in doubt and that it offered no positive benefit to children since there was no evidence that children appreciated the taste. Indeed, the chief reason for adding MSG to baby foods seems to have been to tempt the palate of the mothers who were making purchasing decisions.[53]

The allegations against MSG were aired at hearings before Senator George McGovern's Select Committee on Nutrition and Human Needs and were given substantial television exposure, a circumstance which caused great consternation in the food industry. Bowing to public pressure, three major producers—Gerber Products Company, H. J. Heinz Company, and Squibb-Beechnut, Incorporated—announced in October 1969 that they had stopped using MSG in their baby food products. But at the same time the industry launched a campaign to enlist the Academy's Food Protection Committee as an ally. On October 30, 1969, the food industry, acting through its liaison panel with the Food Protection Committee, made a private plea for Academy support in its battle against its critics. The plea was contained in a letter from Arthur T. Schramm, of Food Materials Corporation, to William J. Darby, then-chairman of the Food Protection Committee. Schramm was serving at the time as head of the industry's liaison group to Darby's committee. His letter is notable both for its tone of panic and its assumption that the Food Protection Committee would unquestionably take up industry's side in the controversy.

"The entire atmosphere growing out of such TV programming, coupled with politically oriented Congressional hearings and careless statements by apparently qualified publicity-seeking individuals, is one of economic terrorism," Schramm wrote.

> Many of the members of the Industrial Liaison Panel, recognizing this sinister development, have expressed strong feelings on the subject and have asked me, as chairman of the Industry Committee, to urge you, as chairman of the Food Protection Committee, to take steps necessary to secure equal time for qualified members of the scientific community to put this matter in proper perspective for the public. While I do not favor the airing of scientific disagreements in public, I feel we have no choice in this particular case, if we are to avoid the type of blackmail that is now occurring under the guise of freedom of speech.
>
> I trust that you will take appropriate action.[54]

What "appropriate" action could be taken to counter the "economic terrorism," the "sinister" developments, and the "blackmail" which Schramm saw threatening the food industry? An opportunity presented itself when the FDA asked the Academy to evaluate MSG and other baby food additives that had come under fire. The Food Protection Committee promptly appointed a special seven-man subcommittee to evaluate the evidence. It was a group that undoubtedly brought a sense of comfort to the beleaguered industry. At first glance, the committee does not appear unbalanced. Two of the members were employed by major chemical companies and thus might be suspected of having an industrial viewpoint on the general question of chemical additives, but their companies—Dow and DuPont—were not directly involved with the production or marketing of MSG. Most of the remaining members were seemingly neutral academics.* But inspection of scientific papers pub-

---

*The panel included Lloyd J. Filer, Jr., University of Iowa pediatrician, as chairman; Pierre M. Dreyfus, chairman of the department of neurology at the University of California at Davis; Kenneth P. DuBois, director of the toxicology laboratory, University of Chicago; Lloyd W. Hazleton, founder of

lished by these members indicates that at least three of them, including Lloyd J. Filer, Jr., chairman of the panel, received research grants from International Minerals and Chemical Corporation (IMC), the major producer of MSG, or Gerber Products Company, a major user of MSG, just before or during the period that the subcommittee was functioning.

Filer, who is Mead Johnson professor of pediatrics at the University of Iowa and former medical director of Ross Laboratories, a manufacturer of infant nutritional products, was involved in a series of studies supported by both IMC and Gerber. He told us that, while the Academy panel was in the midst of its deliberations, members of his research team received grants from the two companies to support studies on MSG. These studies generally exonerated MSG as a hazard. One of them was actually cited in the Academy report as evidence that MSG does not adversely affect the breast milk of lactating women.[55] Two other members of the Academy panel also had ties to the interested companies. Lloyd W. Hazleton's commercial laboratory, Hazleton Laboratories, Incorporated, had been hired by IMC in 1966 to conduct reproduction studies in rats—the studies, which were also cited in the Academy report, exonerated IMC.[56] And George M. Owen, professor of pediatrics at the Children's Hospital in Columbus, Ohio, published a paper in 1969 that was entirely supported by Gerber.[57] (The subject involved cow's milk for infant formulas, not MSG.)

The point about these relationships is not that any of these men were being "paid off" by the industry to return a favorable verdict. Rather, the point is that Filer and Hazleton could be considered partisans on the MSG issue in that their laboratories had done work, under industry sponsorship, which concluded that MSG was safe. Yet they were sitting on a committee that was supposed to balance the studies, including their own, which exonerated MSG against the studies which indicted it as a health hazard. Filer defends these relationships by pointing out that the committee needed knowledgeable experts and that there are

Hazleton Laboratories, Incorporated; George M. Owen, professor of pediatrics, Children's Hospital, Columbus, Ohio; Virgil B. Robinson, Dow Chemical Company; and John A. Zapp, Jr., director of the DuPont Company's Haskell Laboratory.

few proficient food scientists who have not, at one time or other, received research support from the food industry. However, even the Academy's own staff later acknowledged that Filer should probably not have been asked to serve as chairman of the panel, though it defended Filer's probity and saw no reason why he should not serve as a member of the committee.[58]

Filer's subcommittee reviewed virtually every article on MSG in the world's literature with the help of an exhaustive survey that had been compiled by IMC; it sent members to the three main laboratories which had investigated the impact of MSG on the brain and other organs; and, in July 1970, it issued its final report. The thrust of the report was that MSG is safe but that it should nevertheless be removed from baby foods because there is no evidence that it makes foods more acceptable to infants. As the report phrased it, "The Committee concludes that the risk associated with using MSG in foods for infants is extremely small. The committee cannot find, however, that the usage confers *any* benefit to the child and therefore recommends that MSG not be added to food specifically designated for infants." The committee also found "no evidence of hazard from the reasonable use of MSG in foods for older children and adults, except for those who are individually sensitive to the substance."[59]

Since the baby food manufacturers had already stopped using MSG in their products, the Academy report posed no immediate threat to their markets. But it did give them support in their contentions that MSG was a safe substance that had been forced off the market by ill-informed public pressure. What the report failed to note, however, was that the member of the subcommittee who was in the best position to evaluate the crucial brain experiments performed by Olney disagreed with the contention that MSG is safe. He was Pierre Dreyfus, chairman of the department of neurology at the medical school of the University of California at Davis. Dreyfus was the most expert member of the subcommittee in neuropathology, the discipline most relevant to assessing the brain damage caused by MSG. Indeed, after Olney charged that the panel had "poor qualifications to judge the issue,"[60] the Academy staff replied that all members were experts in safety evaluation and that "there was an eminent neuropathologist on the panel in the person of Dr. Dreyfus."[61] Dreyfus was specifically assigned, along with a veterinary pathologist from Dow Chemical, to visit Olney's laboratory. Dreyfus told us that, while the dosages of MSG that Olney administered to experimental animals were higher than the amount likely to be ingested by an infant, he nevertheless felt Olney's experiments indicated that MSG is a risky substance, and he said so in a report to the subcommittee. "I said that a warning should be given that MSG could possibly have some harmful effect," Dreyfus recalls. "I'm not convinced that it really does a great deal of harm but I'm sufficiently concerned about it that I think it should be removed. In other words, I would inject a certain warning, a certain amount of caution." Dreyfus added that he is "convinced . . . that John Olney's findings are correct. These other people [the other members of the subcommittee] are not quite sure of his findings. But I've been there. I've seen the way he works. He's a careful and dedicated guy. He's done good work." In contrast, says Dreyfus, some other laboratories that have done work exonerating MSG are "of questionable reputation."

So why didn't Dreyfus enter a dissenting opinion and say that MSG might indeed be harmful? Because the other panel members "said that based on that visit alone one really could not permit oneself to make an interpretation," Dreyfus recalls. Moreover, Dreyfus did not regard the issue as of much importance once the baby food companies had voluntarily removed MSG.

Dreyfus says he was "sort of the black sheep" on the committee since he was "more critical" than the others. "I think they regretted afterwards that I was asked to serve," he adds.

Olney's findings were contradicted by studies carried out at the other two laboratories that were visited by subcommittee members—the Food and Drug Research Laboratories, a commercial operation, and the Institute of Experimental Pathology and Toxicology at Albany Medical College, Albany, New York. Both laboratories have received substantial financial support from the food industry over the years. And both laboratories have close ties to the Academy's Food Protection Committee. Food and Drug Research Laboratories is headed by Bernard L. Oser, who was a member of the Food Protection Committee at the time the MSG study was conducted. The Albany Medical College unit is headed by Leon Golberg, also a member of the Food Protection Committee at that time. Thus, it is not surprising to find that a "clubby" atmosphere prevailed when the Food Protection Committee's MSG panel called in investigators from these two laboratories for formal presentations of their findings. Listen to Olney describe the proceedings on the day he testified before the panel along with Oser and Frederick Coulston, the chief investigator of MSG at Albany Medical (whose experiments, incidentally, were supported by IMC):

"When Coulston entered the room everyone got up and raced over to greet 'Old Fred.' After much back slapping between Fred and the committee members (Dreyfus was absent that day), Fred strode across the room, sat down at the end of the table as if he were in charge of the session and commenced telling the committee that he fully sympathized with their chagrin at having to waste their time on this nuisance issue. He said that his laboratory had produced the 'so-called brain lesions' in infant mouse brain with MSG but that he could not see why the phenomenon would be called brain damage. He babbled on that he had known of a man being shot through the head with a pistol and still being able to walk and act like nothing had happened so he could not see why anyone should get exercised about the loss of a few hypothalamic nerve cells in infant brain. When Coulston presented some data on mice . . . I tried to pin him down with a series of specific questions about technique and he refused to answer except in terms of 'Don't worry, my people are highly skilled.' Oser described for the subcommittee the protocol he intended to follow in studying MSG effects on mice, rats, dogs and monkeys. . . . I told him in front of the committee that his efforts would be worthless and would only confuse the issue unless a number of specific changes in research design were made. He thanked me courteously for the helpful suggestions, then went on to perform the studies in the same . . . manner described in his original protocol."[62]

Coulston later charged that Olney was guilty of "misrepresentations" when he suggested that the Oser and Coulston studies and the Food Protection Committee review of them were "an industry-arranged whitewash affair."[63] But the inbred atmosphere Olney described raises serious questions as to how objective the Academy committee could be. In the MSG case, there was profound disagreement between Olney, on the one hand, and Oser and Coulston on the other, with the two sides coming up with opposing findings in laboratory experiments. Dreyfus, the only member of the subcommittee who had never served with the Food Protection Committee or other Academy food panels before, sided with Olney. The other members sided with Oser and Coulston. Was this a purely professional judgment, or did personal ties lead the committee to side with the members of its own clique against the crusading outsider?

The food industry was pleased with the findings of the Academy report (although it would have been even happier with a report that concluded MSG was positively beneficial and should be put back into baby food). On October 23, 1970, Dan Gerber, chairman of the board of Gerber Products Company, wrote to doctors around the country assuring them that baby food is safe and nutritious. A leaflet accompanying the letter defended MSG by asserting: "The National Academy of Sciences ad hoc committee in a 42-page report to the Food and Drug Administration pointed out that there is no evidence of any harm in feeding foods to infants containing MSG and that the chances of any such harm are extremely remote."[64] Similarly, sources at IMC assured reporters: "The National Academy of Sciences has given MSG a clean bill of health and the scare has ended."[65]

The World Health Organization was somewhat more cautious, however. In early 1972 the joint FAO–WHO Committee of Experts on Food Additives stated: "In view of the uncertainty regarding the possible susceptibility of the very early human neonate to higher oral intake of glutamate, it would be prudent not to add monosodium glutamate to food specifically intended for infants under one year of age."[66] That was a position essentially identical to the stand Dreyfus had taken earlier in opposing the Academy committee.

Dreyfus had decided not to make a big fight over the safety of MSG in the belief that the question was moot after the baby food manufacturers voluntarily stopped using the substance. But this decision may have been shortsighted. For one thing, young children can still be exposed to MSG through adult foods and soups. No one seems to know how much adult food is fed to small children, but congressional hearings indicated that Campbell Soup Company scientists had told outsiders that "a lot of their foods were now being given to babies, especially soup products."[67] Moreover, there is nothing to prevent the manufacturers from putting MSG back into their baby food products. In mid-July 1971, in fact, IMC asked the FDA to reaffirm the status of MSG as a substance "generally recognized as safe" and argued against putting any limits on MSG in baby foods.[68] IMC subsequently sold its MSG business; and the FDA in late 1974 awaited further safety reviews before deciding whether to impose restrictions on MSG.

The Academy's 1972 evaluation of the hazards associated with a widely used red food dye provides another illustration of how the food industry can look to the Food Protection Committee to protect its interests. The industry pressured the Academy leadership into authorizing a study of the controversial food color known as FD&C Red No. 2 in an effort to head off a regulatory crackdown; the Food Protection Committee came up with a report exonerating the chemical; and the Academy transmitted the report to the FDA despite a blistering internal critique by the Report Review Committee.

Red 2 has been the most widely used of all food colorings. Alone or in combination, it is found in soft drinks, sausages, hot dog casings, ice cream, breakfast cereal, bakery products, gelatins, snack foods, candies, pistachio nuts, pet foods, cosmetics and ingested drugs, among other products. According to one industry estimate in 1972, about $19 million worth of Red 2 is produced annually and is used in $15 to $25 billion worth of food.[69] Says Keith Heine, chief of the colors unit at FDA: "Red 2 is so ubiquitous that, if every food with Red 2 self-destructed tomorrow, a lot of people would starve."[70]

Red 2 has been used in this country since the turn of the century without any serious challenge to its safety. But in 1970 two reports appeared in the Russian scientific literature which suggested that Red 2 might be a hazard to human health. One of these reports purported to find carcinogenic effects. Rats fed a diet containing Red 2 developed significantly more tumors than did a control group which was not exposed to Red 2. The other Russian report claimed that Red 2 caused reproductive failures. Rats fed Red 2 showed a significant decrease in female fertility and a significant increase in stillbirths and deformed fetuses compared to a control group.[71] The FDA, and some public interest scientists as well, discounted the significance of the Russian cancer findings because of weaknesses in the experiment and because apparently sound experiments conducted elsewhere indicated that Red 2 was not a cancer threat.[72] But the FDA had no data to compare with the Russian reproductive findings, so the agency launched several studies, some by its own scientists and some by outside contractors, which seemed to confirm the Russian results. One FDA study found that Red 2 caused fetal deaths (resorption of implanted fetuses) in rats at moderate dosage levels of 30 mg/kg/day with a possible effect at 15 mg/kg/day. A second FDA study found that Red 2 was toxic to chick embryos at varying dosage levels (that was the work of Jacqueline Verrett which the committee chairman had found so "entertaining"). And a third study—conducted by Stanford Research Institute under contract with the FDA— indicated that Red 2 caused genetic damage, though these results were later discounted as the result of impurities in the bacteria used in the experiment.[73] The FDA announced in September 1971 that it might be necessary to set limits to consumption of Red 2 and to allocate the amount of Red 2 that could be used for various purposes.[74] But in November 1971, FDA scientists who reviewed the latest evidence recommended even stronger action—a virtual ban on Red 2 in food products except for "indirect or incidental" applications such as food packaging.[75]

Before launching the far-reaching regulatory crackdown recommended by its own scientists, however, the FDA decided to seek an opinion from the Academy. Much to the FDA's surprise and dismay, the Academy at first declined to consider the matter because of the "routine character" of the questions involving Red 2.[76] In other words, there was no reason why FDA could not determine the safety of Red 2 itself without hiding behind the prestige of the Academy. But the Academy was soon pressured into changing its mind. At a December 1971 meeting of the Food Protection Committee President Handler appeared before the committee's industry liaison group, heard the industry's complaints, and agreed to ask the Academy Council to reconsider its decision. Subsequently the Academy informed FDA it would indeed review the Red 2 issue.

The rationale for accepting the task, as explained in a letter from Handler to two of Ralph Nader's associates, was that the project was expanded to include commentary on FDA's proposed policy of allocating usages, making the Academy review "not necessarily routine."[77] But the report later issued by the Academy barely mentioned the proposed policy of allocating uses; it simply performed the routine safety evaluation which the FDA had originally requested and which the Academy had originally turned down. What seems to have happened is that the Academy was under pressure to do the job from the industry and from FDA, so it knuckled under and performed a task it didn't want to do. "There was some reaction against taking the assignment," says Coon, "but the Academy and the Food Protection Committee finally gave in. It was the industry urging us to do it that I guess finally ended up in our doing it."

Later, Handler acknowledged to us that he may have been "woefully naïve" in agreeing to accept the Red 2 study. He explained that he had walked into the meeting of the Food Protection Committee and its industry liaison panel simply to "say good morning" and was "quite taken aback" when many of those present, most of whom he could not identify, urged him to approve a study of Red 2 and of the philosophy of allocating usages. "It was the second point that I heard that morning really," he said. "You have to understand that at the time I didn't know a damn thing about all this. I didn't know what the problem was with Red Dye No. 2. Nobody had ever told me what had given the FDA cause for concern. I was an innocent.... And if there were any political considerations surrounding it, nobody had ever said anything about those to me.... What I heard was that this is a new way to deal with the problem and they wanted to deal with that."

The Academy appointed a six-man subcommittee, chaired by Coon, to evaluate the evidence. The subcommittee met on February 10, 1972, to hear testimony from FDA scientists, industry researchers and commercial laboratories. It then reconvened the next day and wrote the first draft of a report that exonerated Red 2. This draft triggered a sharp internal dispute within the Academy—one which kept the report bottled up for months while the FDA got increasingly impatient and threatened to take action without waiting for the report. The trouble started the moment the Food Protection Committee sent the report to the Report Review Committee for internal review. At first the authoring committee merely sent its conclusions to the reviewers without submitting the back-up data on which the conclusions were supposedly based. But the reviewers demanded the supporting data, and, when they pored over that data, some of them concluded that it did not support the committee's sweeping conclusions. "The reviewers came down very hard on the report," recalls one Academician. "The Food Protection Committee had concluded that there was no evidence that Red 2 was unsafe. But their report didn't find clear fault with the Russian data and the FDA studies. So how could they say there was convincing evidence of safety?" The dispute between the authoring committee and the reviewers stemmed largely from disagreement over how much weight to attach to various experimental results. The authoring committee put its faith in tests—mainly conducted by commercial and industrial laboratories—which suggested that Red 2 was not a hazard. The reviewers were more impressed by the FDA and Russian findings.

Neither side would back down, so the issue was referred to Handler for resolution. Handler asked the authoring committee to rewrite its report, and when that effort wasn't satisfactory he rewrote it himself, then let the authoring committee revise his revision. But the dispute was never fully resolved, and, when the committee's report was finally forwarded to the FDA in June 1972, it reflected the views of the authoring committee more than those of the reviewers.

The report gave the industry precisely what it wanted. The committee concluded that "there is insufficient reason, today, to take measures to reduce the present extent of human exposure to Red 2, a coloring agent that has been in widespread use since the early days of this century without suggestion of harmful effect on human health." It discounted the three sets of observations which had generated the most concern, namely those involving reproduction, mutagenesis and teratogenesis as "inconclusive." The committee also argued that FDA's plan to allocate usage patterns was "a premature and unnecessary measure at this time. Restriction of general use by some appropriate means can be effected at a later date if future findings warrant such action."[78]

The report was accompanied by a cover letter from Handler which suggested that it should be read as the opinion of one group of experienced scientists rather than as a definitive answer on the safety of Red 2. "It will be evident that the situation has not been adequately resolved," he wrote. "... Decision at this time, therefore, remains a matter of professional judgment which must rest upon insufficient evidence. The judgment of our subcommittee is presented for your consideration."[79] Handler later contended that this cover letter alerted FDA that restrictive action was "entirely in order."[80] But FDA did not show much zeal for a crackdown. The agency announced, on July 3, 1972, only that it proposed to restrict the use of Red 2 to certain levels in food, ingested drugs, lipstick, and pet foods.[81] But as of late 1974, the FDA had not taken even this action, and Red 2 continued in widespread usage pending analysis of further safety tests. Consumer groups and some government scientists were contending that the dye might cause cancer and fetal deaths, while industry and other government scientists poohpoohed the hazard. But whatever the outcome of the debate, the Food Protection Committee's evaluation of the early data revealed that it was still loathe to crack down on possibly hazardous chemicals.

In a 1972 interview, we asked Handler if it would be fair to say that the food industry, upset at the possibility of a ban on Red 2, had pushed the Academy into doing the job because it felt reasonably certain that the Food Protection Committee would advise against regulatory action unless there was strong evidence of hazard. "I think you paint the situation in firmer and sharper terms than the reality," Handler replied. "But those elements were in the playlet that day."

The Academy has taken some steps to "sanitize" the relationship between the committee and industry. In 1972 it attached the industry liaison panel to the committee's parent unit, the Food and Nutrition Board; industry contributions will henceforth be funneled into the parent board for redistribution rather than directly into the committee. But this change appears to be more cosmetic than substantive. The Food and Nutrition Board is not, after all, very far removed from the Food Protection Committee. The chairman of the Food and Nutrition Board in 1973–1974, for example, was Lloyd J. Filer, Jr., who had previously headed the Food Protection Committee and had directed that committee's MSG report; the executive secretary of the Food and Nutrition Board is now Paul Johnson, who had previously served in the same capacity for the Food Protection Committee for almost two decades. Moreover, industry will still retain a voice in Food Protection Committee operations through a small industry liaison panel that will continue to serve as a point of contact though not of direct funding.

To put more balance into the deliberations of the Food Protection Committee, the industry input should be offset by a strong, qualified consumer voice, broadly representing the public interest. In late 1974, under pressure from within the Academy and without, the Food and Nutrition Board asked consumer groups to set up such a panel. But lack of funds may limit the panel's effectiveness. The Academy declined to underwrite the venture, and few consumer groups felt able to contribute much financial support. If no funding is found, the Academy's food liaison panels will probably continue to be dominated by scientists from well-heeled corporations that are happy to pay for the privilege of presenting their views to the Academy's food committees.

The Food Protection Committee's financial dependence on the food and chemical industries remains a serious problem. Perhaps industry funding should be done away with altogether. The industry funding is not overwhelmingly large. And if the price of that support is that the Food Protection Committee thinks of itself as an appendage of industry, then that price is too high. If the committee's work has been defective and biased, as this chapter suggests, then perhaps the nation would be better served if the work were not done at all unless it can be done under more objective circumstances. Who has really benefited from the Food Protection Committee's reports—the public, or those industrial interests that hope to see their viewpoints given stature by the distinguished imprimatur of the Academy?

CHAPTER NINE

Pesticides:
The Academy versus Rachel Carson

"The Academy has been completely incapable of dealing with the pesticides problem because it's been dominated by chemists and because it has never had and still does not have an ecosystem point of view."

> —Roland C. Clement, vice-president, National Audubon Society, in an interview, January 4, 1972.

"The predicted death or blinding by parathion of dozens of Americans last summer must rest on the consciences of every car owner whose bumper sticker urged a total ban on DDT."

> —Philip Handler, president of the National Academy of Sciences, in an address delivered on December 26, 1970, at the annual meeting of the American Association for the Advancement of Science.

"What does it mean," Rachel Carson once asked, "when we see a committee set up to make a supposedly impartial review of a situation, and then discover that the committee is affiliated with the very industry whose profits are at stake?"[1] Ms. Carson was complaining about the Academy's Committee on Pest Control and Wildlife Relationships, which convened in the early 1960s to assess the impact of pesticides on wildlife. That particular committee had indeed fallen under the control of the very agricultural and industrial interests which were promoting the use of pesticides. As a result, at the very time that Ms. Carson's best-selling exposé, *Silent Spring*, was first alerting a wide audience to the damaging side effects of pesticides, the Academy was leaning its authority in the opposite direction. The Academy committee became a focal point for counterattacks against Ms. Carson and its reports, she lamented, were "frequently cited by the pesticide industry in attempts to refute my statements."[2]

The Academy's total contribution to the decade-long pesticides controversy is difficult to characterize. A succession of Academy committees has issued reports on aspects of the problem since the early 1960s. Some, such as the reports that annoyed Ms. Carson, have been unabashedly pro-pesticides; some have been so bland that they defy characterization; and at least two have been highly critical of pesticide abuses. But, on balance, it seems fair to say that the Academy has seldom, if ever, been in the vanguard of those seeking to curb the excesses of pesticide pollution. Indeed, the Academy's most comprehensive pesticide reports have brought comfort to the forces of agriculture and industry while they have caused dismay among conservationists. Compared with reports issued by other prestigious scientific groups—notably the President's Science Advisory Committee and the Secretary's Commission on Pesticides and Their Relationship to Environmental Health (the Mrak Commission), a group of distinguished scientists appointed by the Secretary of Health, Education, and Welfare—the Academy's reports were weak and equivocating. There is even evidence, in the form of internal documents which have been made available to us, that the Academy's pro-pesticides bias has provoked criticism from environmentalists operating within the Academy structure.

The recent widespread concern over chemical pesticides stems largely from the publication in 1962 of Ms. Carson's remarkable book, *Silent Spring*. Ms. Carson's purpose in writing the book was avowedly to draw attention to the detrimental side effects of pesticides. Most previous discussions of pesticides had stressed their beneficial aspects, particularly their undeniable value in controlling insects which would otherwise destroy crops, damage forest lands or spread disease. Ms. Carson, on the other hand, consciously set out to dramatize the other side: the harmful side effects that occur when potent chemicals are spread indiscriminately throughout the environment—particularly the damage done to beneficial insects and wildlife and the potential health danger posed to man himself. Ms. Carson did not, it should be emphasized, call for an end to all use of pesticides. She directed her fire chiefly at certain persistent, or long-lasting, pesticides which tended to spread widely throughout the environment and whose residues were concentrated in animal tissues. For shorter-lived pesticides, she counseled restraint. And she repeatedly urged greater use of non-chemical control techniques.

Ms. Carson's book was subjected to bitter attacks from the chemical industry, but the next year her concerns were essentially endorsed by the President's Science Advisory Committee (PSAC). "Until the publication of 'Silent spring' by Rachel Carson, people were generally unaware of the toxicity of pesticides," PSAC said. "The Government should present this information to the public in a way that will make it aware of the dangers while recognizing the value of pesticides."[3]

The reaction at the Academy was much frostier, however. Key members of the Academy's Committee on Pest Control and Wildlife Relationships were among Ms. Carson's most vociferous critics in the scientific community. In a review of *Silent Spring* written for *Chemical World News*, committee member George C. Decker, an economic entomologist with the Illinois Natural History Survey who had frequently served as a consultant to the chemical industry, stated: "I regard it as science fiction, to be read in the same way that the TV program 'Twilight Zone' is to be watched."[4] In hearings on Capitol Hill, committee member Mitchell R. Zavon, a consultant for Shell Chemical Company, described Ms. Carson as one of the "peddlers of fear" whose campaign against pesticides would "cut off food for people around the world."[5] And in a long, critical review of *Silent Spring* published in *Science,* committee chairman I. L. Baldwin complained that the book was not "a judicial review or a balancing of the gains and losses; rather it is the prosecuting attorney's impassioned plea for action against the use of these new materials which have received such widespread acceptance, acceptance accorded because of the obvious benefits that their use has conferred." Much sounder information, Baldwin suggested, could be found in the "balanced judgments" of reports issued by his own and other Academy committees.[6]

Ms. Carson was also attacked in scurrilous fashion by two scientists who were not members of the pest control committee but who have long played influential roles in the Academy's food and nutrition activities. One was William J. Darby, a Vanderbilt University nutritionist, who wrote a critical review for *Chemical and Engineering News* in which he accused Ms. Carson of "ignorance or bias" and complained that she had ignored the "sound appraisals" of such "responsible" bodies as the National Academy of Sciences. Darby showed his own bias by misrepresenting Ms. Carson's position (he claimed she said it was neither wise nor responsible to use pesticides in the control of insect-borne diseases) and by suggesting that her book would appeal to "organic gardeners, the anti-fluoride-leaguers, the worshippers of 'natural foods' and other pseudoscientists and faddists."[7] Similar sentiments were expressed by Academy member C. Glen King, acting as head of a nutrition organization not a part of the Academy.[8]

Ms. Carson had some defenders among the members of the Academy, notably Harvard biologist John T. Edsall, who found her philosophy "more objective and judicial than that expressed by some of her critics."[9] But Edsall's comments, it should be noted, were expressed as an individual. Those who held the levers of power on the Academy's pesticide and food committees seemed generally hostile.

How balanced were the Academy's own reviews of the pesticides problem? Let us take a close look at the Committee on Pest Control and Wildlife Relationships—the group which Baldwin said exercised "sound judgment based on facts."[10] The committee was originally formed in 1960 in an effort to cope with some of the same problems that led Rachel Carson to write her book. Financial support came from government and industry sources, with a small amount from conservation groups. But the composition of the committee was decidedly unbalanced. It was dominated by scientists who would tend to support the use of chemical pesticides to control insects as opposed to those who would be sensitive to deleterious side effects.

The project was directed by a seven-man parent committee. Only one member was a wildlife proponent; the others were industrial consultants or came from backgrounds in agriculture, forestry, and entomology, the very fields most identified with use of pesticides.* Probably the most important single influence on the committee was the late W. H. Larrimer, the Academy staff member who served as executive secretary and played a key role in drafting the committee's reports. Larrimer had spent most of his career in the insect control and forestry programs of the Department of Agriculture. He thus had a long professional commitment to the use of pesticides. In the opinion of the late wildlife specialist Clarence Cottam, a vigorous supporter of Rachel

---

*The parent committee was chaired by I. L. Baldwin, former dean of agriculture at the University of Wisconsin, and included: George C. Decker, an economic entomologist and consultant to the chemical industry; Ira N. Gabrielson, president of the Wildlife Management Institute; Tom Gill, head of a forestry foundation; George L. McNew, a specialist on control of crop diseases who was managing director of the Boyce Thompson Institute for Plant Research, a lab with a large industrial clientele; E. C. Young, an agricultural economist at Purdue; and Mitchell R. Zavon, clinical professor of industrial medicine at the University of Cincinnati, who consulted for Shell Chemical Company, a maker of pesticides.

Carson who served on one of the subcommittees, Larrimer was "convinced beyond any question that no ill effects could result from the use of any of these chemicals."[11]

The parent committee appointed three subcommittees, consisting of eight members each (some of whom were also on the parent committee), to write reports on specific topics. The chairman of the first subcommittee, which was given the crucial job of evaluating the extent of pesticide–wildlife problems, was Decker, the economic entomologist who had served as a consultant to industry and had described *Silent Spring* as "science fiction." At least half of the subcommittee members are known to have worked for or consulted with the chemical industry. The second subcommittee—which was to recommend policies and procedures for pest control—was dominated by scientists from industrial, agricultural, or other backgrounds that would predispose them to favor use of pesticides. Only on the third subcommittee, which had the innocuous task of evaluating research needs, were the wildlife interests in a strong position. Considered as a whole, Cottam told the study team, the Academy committee was "a handpicked group. It had a preponderance of pesticide workers from the agricultural or chemical industries. I thought it was collusion."

The committee's first two reports were published in 1962, shortly before *Silent Spring* first appeared. Not surprisingly, they took a rather lenient view of the hazards of pesticides. The first report, entitled "Evaluation of Pesticide–Wildlife Problems," was twenty-eight pages long. After a brief introduction, it opened with a five-page section describing how pesticides are "a modern necessity" in agriculture, forestry, and public health. This was followed by a section on "wildlife values" only two pages and five lines long. This section stressed that, while the Indians once relied on wildlife for their very existence, "today's emphasis is on recreational and esthetic uses."[12] Then followed a long section assessing the impact of pesticides on wildlife. Although pesticide hazards were acknowledged and some losses admitted, the section suggested that such losses have been "minimal" and have usually resulted from careless application. The report almost completely ignored the problem of pesticide concentration in the food chain of various species; it barely mentioned alternate means of control; it made no attempt to specify which pesticides pose hazards and how those hazards might be reduced; and it gave little hint of the problems which led to the formation of the committee in the first place. Like many Academy reports, this one contained essentially no documentation. Its pronouncements, which must be accepted on faith, relied on the Academy imprimatur to give them authority.

The second report, entitled "Policy and Procedures for Pest Control," was essentially a fifty-three-page handbook for operators of pest control programs. Though this report was better balanced than the first, it largely accepted the existing emphasis on chemical pesticides, acknowledging that "The material is directed chiefly to large-scale operations involving the use of chemicals."[13] The report recommended twelve procedures and seven precautions that should be followed in order to carry out an effective and safe pest control program, but it made no effort to evaluate existing pest control programs against these criteria.

These two reports, which supposedly constituted the Academy's best judgment on the problems that were troubling Rachel Carson, were greeted with dismay by many ecologists and wildlife experts. In a critical review of the documents, Roland C. Clement, staff biologist of the National Audubon Society, commented:

> Given their authoritative backing, these two reports are extremely disappointing ... the whole is a generalized and undocumented statement that, far from coming to grips with the problem, seems to disregard much important evidence and does no more than offer a gentle admonition to pesticides users to be more careful in the future. This result is either a mark of ecological incompetence in the committee or, more likely, evidence that the viewpoints of the advocates of pesticides-use within the committee prevailed almost entirely.[14]

Ecologist Frank E. Egler was even more critical. Writing in the *Atlantic Naturalist*, he complained that the first report treated wildlife as "something that annoyingly gets in the way of pest control programs." As Egler saw it, "This booklet 'explains' the status quo, with a minimum amount of factual material, with no documentation or references to the scientific literature, and with absolutely no implication that there is the slightest thing wrong with the status quo." The two reports taken together, he said, "cannot be judged as scientific contributions. They are written in the style of a trained public relations official of industry, out to placate some segments of the public that were causing trouble. It is my opinion that with different title and cover pages, they would serve admirably for publication and distribution by a manufacturers' trade association. Indeed, they are being much quoted in such places."[15]

The first two reports also caused consternation among some members of the third subcommittee, according to internal correspondence which has been made available to us. Clarence Tarzwell, an aquatic biologist who later became director of the National Marine Water Quality Laboratory in West Kingston, Rhode Island, charged that "There has been a constant playing up of the great benefits and need of pesticides and a playing down of harmful effects and the detrimental effects on wildlife."[16] Similarly, Clarence Cottam found "bias and a lack of objectivity." He added: "An important theme throughout all this study has been the defense of chemical pesticides rather than objective control by whatever means becomes necessary.... I have been disturbed because of the efforts to maximize the benefits and minimize the ill effects of the use of chemical pesticides."[17]

Cottam, in particular, became so incensed that he launched a vigorous campaign to make certain the third report did not come out with the same slant as the first two. He was soon involved in constant bickering with Larrimer, the committee's staff man. As Cottam explains it, Larrimer "changed most everything I wrote and several times it came back just a little stronger in the line that he wanted in the first place."[18] Cottam says he tried to get permission to write a minority report but was denied the opportunity. Instead, the subcommittee wrangled on for months before finally agreeing on a brief compromise report that was so innocuous it offended few of the contending interests. The report recommended more research in six general areas but did not specify which pesticides, programs, uses, or wildlife should be studied. "The report wouldn't do credit to a high-school graduate," Cottam says. "It's not worth the paper it's written on." At one point Cottam and Tarzwell tried to get a sentence inserted in the third report saying they did not concur with the first two reports. Such an explicit attack on the credibility of the first two reports could not be tolerated, but the third report does state, in the foreword, that "No subcommittee member, except those who were also members of the main committee, bears any responsibility for the report of any subcommittee other than his own." Thus does the Academy gloss over the deep internal conflicts which might lead to public embarrassment.

Credit for raising the pesticides issue went largely to Rachel Carson. The Academy's initial reaction, expressed through members of the Committee on Pesticide and Wildlife Relationships, had been to pooh-pooh concern over side effects and to issue attacks on Ms. Carson for sounding the alarm. (Years later, a more enlightened Academy committee handed down a telling, if indirect, judgment on the whole affair. In 1970, the Committee on Agricultural Land Use and Wildlife Resources compiled a list of five major investigations into the impact of pesticide use on wildlife. Conspicuously missing was any mention of the three reports that the Academy's own Committee on Pest Control and Wildlife Relationships had issued in 1962–1963.[19])*

The Academy's next major report on pesticides is remarkable for two reasons: It illustrates how government agencies often use the Academy as a shield to head off potential criticism; and it reveals that the Academy's pesticide advisory apparatus was still in the hands of the pesticides advocates.

The stimulus for the report came not from the Academy itself, but from the President's Science Advisory Committee (PSAC). One of the major recommendations of a 1963 PSAC report, *Use of Pesticides*, had called for the Academy to study

---

*The Academy reports look weak not just from the perspective of the conservationists; they were considerably weaker than reports issued by the President's Science Advisory Committee (PSAC), the most prestigious science advisory group in the federal government. A 1963 PSAC report, entitled *Use of Pesticides*, warned that "Many kinds of insect control programs have produced substantial mortalities among birds and other wildlife"; it recommended an "orderly reduction in the use of persistent pesticides ... Elimination of the use of persistent toxic pesticides should be the goal." A 1965 PSAC report, "Restoring the Quality of Our Environment," concluded that "Virtually complete reliance upon chemical pesticides is a mistake. ... Substantial reduction in insecticide use, in specific cases as much as 50%, can be made by applying our present knowledge of pests and their control."

the concepts of "no residue" and "zero tolerance" as they were then used in the registration of pesticides and the setting of tolerances for pesticide residues in food.[20] For years, the Department of Agriculture had routinely registered pesticides on a "no-residue" basis if data submitted by the manufacturer indicated that no detectable residue would remain on food crops as a result of the proposed use. Similarly, the Food and Drug Administration, which was responsible for establishing safe tolerances for pesticide residues in food, had for years been setting some tolerances "at zero level." Sometimes this meant that a pesticide was so toxic that absolutely no residue could be permitted; but often it simply meant that, in practice, no residue was detectable so the tolerance might as well be set at zero. Then, in the early 1960s, problems began to develop. New analytical methods, a thousand times more sensitive than the old, came into use. Whereas the old techniques had detected residues at levels in the parts-per-million range, the new methods could detect some residues down to parts per billion. Suddenly many products previously considered safe were technically in violation of the standards. In an effort to resolve the confusion, the Food and Drug Administration and the Department of Agriculture on June 1, 1964, contracted with the Academy to study the technical issues involved.

The committee that the Academy appointed to examine the problem consisted primarily of scientists from agricultural schools and from the food and chemical industries. The chairman of the thirteen-member group was James H. Jensen, a plant pathologist who was then president of Oregon State University, a strongly agricultural school. The other members included two employees of chemical companies, a chemical engineering consultant, two officials of laboratories serving the food and drug industry, five scientists from agricultural institutions or departments, and two other academics.*

---

*In addition to Jensen, the committee included: Karl H. Beyer, Jr., vice-president, Merck Sharp & Dohme Research Laboratories; W. Donald Cooke, dean of the graduate school at Cornell University; Cuthbert Daniel, chemical engineering statistical consultant; Francis A. Gunther, professor of entomology, University of California at Riverside; T. Roy Hansberry, Shell

To virtually no one's surprise, the committee, in a report issued in June 1965, concluded that "The concepts of 'no residue' and 'zero tolerance' as employed in the registration and regulation of pesticides are scientifically and administratively untenable and should be abandoned."[21] In place of these old categories, the committee recommended a new system whereby pesticides would be registered if their use resulted in "negligible or permissible" intake as determined by appropriate safety studies. On the issue of carcinogenic (cancer-producing) chemicals—the one case where many scientists would argue that zero tolerance should be retained—the committee hedged, asserting that while "It is reasonable to assume that a no-effect level could be demonstrated for a compound with respect to carcinogenic potential, approval of such a compound for use when it might leave a residue on food would require most extraordinary justification."[22]

Not all of the committee's recommendations were accepted by the funding agencies, but the basic thrust of the report seems to have been implemented.[23] Indeed, there is wide agreement on both sides of the pesticides controversy that the government had little choice. The chemical industry trade associations applauded the change. And even Roland Clement, long a battler against pesticide abuses, says: "I went along with this because the new instrumentation denied the significance of the old zero. It's hopeless. The environment is so polluted now that you would have to stop eating everything if you insisted on zero. So it's tragic, but that's the way it is."

Agreement was so general, in fact, that some scientists (including some on the Academy committee itself) have since questioned why the problem was thrown to the Acad-

Development Company; Allen B. Lemmon, California Department of Agriculture; William P. Martin, Institute of Agriculture, University of Minnesota; James A. Miller, professor of oncology, University of Wisconsin; Emil M. Mrak, chancellor, University of California at Davis; Kenneth E. Mulford, Atlas Chemical Industries, Incorporated; Bernard L. Oser, Food and Drug Research Laboratories, Incorporated; and Charles E. Palm, dean, college of agriculture, Cornell University. Staff chores were handled by Harry W. Hays, then director of the Academy's Advisory Center on Toxicology, which has close ties to the chemical industry, while A. Geoffrey Norman, chairman of the Academy's division of biology and agriculture, provided advisory assistance.

emy in the first place. "The federal agencies knew what the answer was before they went to the Academy," asserts Ned D. Bayley, director of science and education for the Department of Agriculture. "They used the prestige of the Academy to convince people and to hold off those who were pushing for retention of the zero-tolerance concept. The government essentially used the prestige of the Academy to confirm its own thoughts." Bayley says that, while he largely agrees with that particular report, such use of the Academy "goes against my craw just a little."

The "zero-tolerance" report is also notable because it backed the status quo in pesticide regulation by advocating that responsibility for registering pesticides should remain solely with the Department of Agriculture, the very agency which critics contended was responsible, in large part, for the existence of a pesticides problem.[24] This suggestion was in sharp contrast to an earlier recommendation by PSAC and a later recommendation by the Mrak Commission that two other departments—Interior and Health, Education, and Welfare—be given a substantial role in approving pesticides.[25] The Academy panel's judgment on this issue was later shown to be faulty by revelations—emanating from the General Accounting Office and the House Government Operations Committee—that Agriculture had consistently failed to protect the public from pesticide hazards.* These

---

*In 1968, the General Accounting Office issued a report highly critical of the Agriculture Department's regulatory enforcement procedures for pesticides; it charged, among other things, that for the previous thirteen years Agriculture had not reported a single alleged violator of the pesticide registration laws to the Department of Justice for prosecution, even though GAO felt prosecution was warranted in some instances, particularly when there had been serious and repeated violations by some offenders.[26] Similarly, in 1969, a House subcommittee held hearings and issued a damning report of its own. The report charged that, until mid-1967, the Agriculture Department's Pesticides Regulation Division (PRD) had "failed almost completely ... to protect the public from hazardous and ineffective pesticide products." It found that hundreds of pesticide products had been approved for registration over objections of the Department of Health, Education, and Welfare as to their safety; that pesticide products had been approved for uses that "were practically certain to result in illegal adulteration of food"; that labels approved for registration failed to inform users of hazards; that PRD had *never* secured cancelation of a registration in a contested case; and that, even in cases where the manufacturer voluntarily agreed to cancelation, the PRD "consistently failed" to take action to remove the hazardous products from marketing channels.[27]

derelictions were occurring at the very time that the Academy was recommending that Agriculture remain in charge of pesticide registration. But what could one expect from a committee composed of agriculturalists and industrialists who submitted their recommendations in advance to Agriculture for comment?[28]

In 1967, in one of those curious about-faces that make it so difficult to generalize about the Academy, a new pesticides committee prepared a hard-hitting report that criticized one of the Department of Agriculture's most controversial programs—the massive campaign to eradicate the imported fire ant. The "ant war," as it is sometimes called, has long been a popular cause with several powerful Southern Congressmen. The fire ant, a tiny dark or red insect with a fiery sting, seems to have entered this country from South America around 1918. Over the next four decades it spread from its original invasion point at Mobile, Alabama, into most of the Southern states. Complaints were occasionally heard that the high, hard mounds built by the ant interfered with haying and mowing operations, or that the painful sting of the insect was a nuisance to farmworkers, picnickers, and even children on urban playgrounds. But it was not until the late 1950s that a concerted effort was launched to eradicate the pest. In 1957, at the instigation of several Southern Congressmen, legislation was passed authorizing the Department of Agriculture to destroy the ant and appropriating funds to begin the campaign, which was to be financed 50–50 by state and federal governments. The Agriculture Department promptly launched a publicity blitz depicting the fire ant as a menace to agriculture and human health; and it announced plans for treating up to 20 million acres in nine Southern states.[29]

Almost immediately the program came under attack from conservationists who charged that massive application of chemicals would cause harm to humans, animals, and other insects and who alleged that the fire ant was not much of a pest anyway. As the criticism mounted, the Agriculture Department first cut down on the amount of heptachlor (a highly toxic, persistent pesticide) it was recommending be used against the fire ant, then abandoned heptachlor entirely and substituted Mirex, which is less persistent, more precise in killing its target organisms, and less toxic to animals. Opposition from conservationists continued, however, on the grounds that Mirex tends to concentrate in the food chain and is a potential carcinogen.

The program was also challenged on economic grounds, with questions being raised about the economic significance of the pest and about the feasibility of achieving the costly goal of eradication. In 1965 the General Accounting Office criticized the program,[30] and for two successive years in the mid-1960s the Budget Bureau tried unsuccessfully to delete funds for the program—only to have Congress put the money back in, with a particularly hefty increase in fiscal 1967.

At this juncture, frustrated budget officials decided to bolster their case against the fire ant campaign with an objective scientific study. Russell McGregor, who was then budget examiner for agriculture, approached the White House Office of Science and Technology but found that "they wouldn't touch it." He then persuaded Agriculture, which had mixed feelings about the utility of the "ant war," to commission a study by the Academy.

The committee which the Academy put together to evaluate the program must, in all candor, be described as "loaded" against the fire ant campaign. Although some previous Academy committees were dominated by chemical control advocates, this particular committee seems to have deliberately excluded anyone directly involved in fire ant control. The chairman of the committee was the late Harlow B. Mills, an entomologist who had conducted a previous evaluation of the program in 1958 under contract with the departments of Agriculture and Interior. At that time Mills, in a report that was never made public, had concluded that the goal of eradication was unrealistic and the cost exorbitant and that even control of the ants, as opposed to eradication, was not a proper federal activity.[31] At least two other members of the twelve-man committee had also previously attacked the fire ant program. Both F. S. Arant, head of the department of zoology and entomology at Auburn University, and L. D. Newsom, head of the department of entomology at Louisiana State University, had been quoted as critical of the program in *Silent Spring*.[32]

The committee held three meetings, toured the South for

on-site inspections, heard testimony from experts and interested parties, and even made a point of having each member stung by fire ants to experience the pain. On September 28, 1967, it submitted its final report to Agriculture. The conclusions were a sweeping indictment of the fire ant campaign. The committee said it felt that "eradication of the Imported Fire Ant is not now biologically and technically feasible"; and it expressed "grave doubts whether an attempt to eradicate it would be justified, even if it were shown to be feasible at a later date." The committee said it was unlikely the ant would spread throughout substantial portions of the United States as the Agriculture Department had been contending; it found no evidence that the ant was a major public health problem, a menace to fish and wildlife, or a serious hindrance to agriculture; and it disputed virtually every justification that had been put forth in support of the program. The committee acknowledged that the fire ant was "an extremely irksome pest" and that its venom posed a danger to extremely allergic persons, but it rated the ant "a nuisance, ranking in importance below other biting and stinging insects." The ant could be kept at tolerable levels, the committee suggested, by local control measures where needed rather than by a massive eradication campaign.[33]

No more damning indictment of a federal pesticide program had ever been prepared by the Academy. The only trouble was that the report was not made public. There was a fleeting reference to the Academy's negative findings during 1968 hearings on the Agriculture Department's proposed budget for fiscal 1969. But Congressman Jamie L. Whitten (D-Miss.), chairman of the House Appropriations Subcommittee for Agriculture, and a champion of the fire ant program, quickly cut off further discussion and told the Agriculture Department to "be sure to enlarge your statement to carry the full story."[34] No copy of the report was ever published by the subcommittee; the Agriculture Department distributed only about twenty-five copies;[35] and the Academy, deferring to Agriculture, steadfastly refused to make known its views. The report was not publicly available until Senator Gaylord Nelson obtained a copy and inserted it in the *Congressional Record* on July 15, 1971—four years after it had been completed.

The Academy's suppressed report had little impact on the

fire ant program. The acreage treated annually reached an all-time high of 20 million acres in 1973 and was projected to reach even higher in 1974. Meanwhile, annual funding for the program topped $7 million for the first time during the 1970s. By the end of fiscal 1973, the Agriculture Department had obligated more than $60 million to fire ant treatment since the program began in 1958, and state governments had contributed more than $40 million in matching funds—a total of more than $100 million spent for a program which much of the scientific community deems unwise and even destructive.[36] The campaign rolled on with the strong backing of key Southern congressmen and the commissioners of agriculture in several Southern states. The "ant war" apparently owes its political strength to the fact that it enables politicians to show that they are doing something for their people, and it provides a convenient way to funnel federal funds into the states.[37] The only factors which have slowed the growth of the program in recent years have been budget stringencies which forced some states to cut back on their matching funds, and a decision by the Environmental Protection Agency in 1972 to prohibit aerial spraying of Mirex in coastal counties, aquatic areas, and heavy forests.

In such a tangle of powerful political forces, it would perhaps be too much to expect a report from the Academy to exert much influence. But the Academy surely did nothing to rectify a policy it considered misguided when it acquiesced in the virtual suppression of its unusually hard-hitting report.

Two years after completing its tough fire ant report, the Academy flip-flopped again and published a report on persistent pesticides that brought joy to the chemical industry but was considered weak and pollyannaish by conservationists. It was also criticized by the Academy's own Environmental Studies Board, and even by the Agriculture Department, which could hardly be considered an irrational or implacable foe of pesticides. Most of the minutes and internal correspondence relating to the report have been made available to us, so this project provides an unusually well-documented opportunity to assess the factors that led an Academy committee to fail in its assigned task.

The origins of this project lay in a letter, dated November 29, 1966, from George Mehren, then Assistant Secretary of Agriculture, to Frederick Seitz, then president of the Academy. Mehren noted that the 1963 PSAC report had called for an "orderly reduction in the use of persistent pesticides" and had further recommended that "elimination of the use of persistent toxic pesticides should be the goal." He expressed hope that the Academy would produce "recommendations and guidelines" that would help Agriculture make "any justifiable changes" in its policies toward pesticides.[38] Subsequently, a contract was signed which called for the Academy to establish "criteria" by which Agriculture could evaluate its pesticide policies "with a view to their appropriate modification or discontinuation." The contract also called for the Academy "to evaluate the possibilities for substituting available non-persistent pesticides for persistent chemicals now being recommended."[39] The Agriculture Department, which had been under constant fire for allegedly reckless promotion of pesticide usage, seems genuinely to have wanted advice on how best to cut back on use of persistent chemicals.

Unfortunately, the Academy was not the place to get such advice. The project was assigned to an ad hoc committee, known as the Committee on Persistent Pesticides, which was headed by James H. Jensen—the same Jensen who had headed the 1965 report which endorsed the status quo in pesticide registration at the very time the whole registration system was falling into disrepute. The new Jensen committee labored over its assignment for roughly a year and a half before finally producing a report in May 1969 that can charitably be described as ducking all the main issues—and it might even be considered a whitewash.

The thirty-four-page report starts out by quoting an excerpt from the 1963 PSAC report which suggested that modern agriculture "necessitates the use of pesticides with their concomitant hazards." At the outset it thus sets a positive tone toward pesticides even though the thrust of that PSAC report had been an endorsement of the concerns over pesticide pollution. The Academy report then goes on to a vague discussion of the uses and effects of pesticides. It ends with eighteen conclusions and seven recommendations, not one of which is truly responsive to Agriculture's major questions. The chief recommendation is that "further and more effective steps be taken to reduce the needless or inadvertent release of persistent pesticides into the environment"—an exhortation that would win unanimous agreement from all sides of the pesticide controversy but which included no hint as to what uses might be considered needless.

The report makes no effort to provide "criteria" by which Agriculture could evaluate its pesticide policies despite Agriculture's specific request for such guidelines. And not a single recommendation deals with the question Agriculture had asked concerning the possibility of substituting available nonpersistent chemicals for persistent ones; one of the conclusions does state that persistent pesticides are "essential in certain situations," but it does not spell out which pesticides and what situations.[40]

Amazingly, the Academy report recommended that "The present system of regulation, inspection and monitoring to protect man and his food supply from pesticide contamination be continued."[41] The committee was apparently oblivious of, or chose to ignore, the facts that were beginning to emerge concerning Agriculture's maladministration of pesticide regulation. As we have seen, a 1965 Academy committee headed by Jensen had similarly endorsed the status quo in regulation. In the interim, the General Accounting Office had issued a report, in 1968, that was highly critical of the Agriculture Department's pesticide regulation and a Congressional committee was gearing up for hearings that would expose the pesticide regulation program as a near-total failure. Yet the Academy committee continued to urge that the existing system be retained.

As it turned out, little attention was paid to the Academy's advice on that issue. In 1970, responsibility for regulating pesticides was shifted from Agriculture to the new Environmental Protection Agency in recognition of the fact that pesticide pollution had become too big a problem to be left to the pesticide promoters.

Even the Agriculture Department, long considered a promoter of pesticides, was dismayed at the quality of the Academy's report. "The PSAC committee had said that we should reduce the use of persistent pesticides," laments Ned D. Bayley, Agriculture's top science administrator. "A lot of people were reaching for just exactly how to go about doing that and wanted to have Academy recommendations to lean on so they could do battle with the industry on it. We didn't get them. It left us kind of high and dry." Bayley complains that the Academy's "general recommendations to restrict the use of persistent pesticides" were "not too helpful" because "we'd been going in that direction all the time—we wanted to know what chemicals, what timetable."

The Academy's own Environmental Studies Board, an interdisciplinary group which seeks to coordinate the Academy's pollution studies, was disturbed at the report's apparent bias. When the board was shown a draft copy of the report's conclusions and recommendations on February 20, 1969, it made a number of comments, most of which "were critical of the conclusions and recommendations," according to an internal memorandum. The board was particularly concerned that the first conclusion of the report, as then worded, "gives the impression that persistent pesticides are beneficial and that no action should be taken in regard to them." The board then went on to complain that the conclusions, taken as a whole, were not strong enough. The board also said the report "should consider the basic question of whether the government has the right to subject the world to long-term effects of pesticides when there is no understanding of long-term effects."[42] So far as can be ascertained from the internal documents available to us, the comments of the Environmental Studies Board had relatively little effect on the final form of the report.

The report, not surprisingly, was considered a boon by industry. *Industrial Research* described the report as "a whitewashing of 'hard' pesticides" which "gave momentary comfort to the storm-tossed industry."[43] *Chemical Week* agreed that the report "should bring comfort to the agricultural chemicals industry."[44] And the report was used to support the chemical industry position in DDT hearings in Washington in 1969 and before the Nebraska legislature in 1970.[45]

Conservationists, on the other hand, were sharply critical of the report. Charles F. Wurster, chairman of the scientists' advisory committee of the Environmental Defense fund, complained in a letter to Senator Philip A. Hart, who was holding pesticides hearings, that "the benefits of persistent pesticides were overstated, and that deleterious effects were understated."[46] The Audubon Society's Roland Clement was only slightly less harsh. In a review entitled "The Academy of Sciences Lays Another Thin-Shelled Egg," he found that this latest report marked only a small advance over previous Academy efforts:

> Though its recommendations are altogether vacuous, it does recognize the damning evidence—but carefully buries it deep in the text, where few will read it. More research is called for, of course, which is always a good way to put off action. . . . The other reliable roadblock to restrictive action is to insist, as this report does, that there are yet no satisfactory alternatives to DDT. It does not specify what it is that cannot be controlled by other means, or how important it is to control at all. . . . We have a right to expect more from the National Academy of Sciences. . . . Someone has to lead, to educate to hold the mirror up to man.[47]

The Academy report looks especially weak when compared with a report prepared a short time later for the Secretary of Health, Education, and Welfare by a special commission headed by Emil M. Mrak, chancellor emeritus of the University of California at Davis. The Mrak Report is the most thorough and extensively documented report on pesticides ever issued by the federal government. It is 677 pages long (including subpanel reports and citations), compared with only thirty-four pages for the Academy report. It reviews over five thousand references—whereas the Academy reviewed only about 285. And it names 138 experts as "a partial list" of those consulted; the Academy consulted only eighty-three. More important than mere statistics, however, is the style and tone of the report. Whereas the Academy report was weak and vague, the Mrak Report was tough and specific. It recommended that all uses of DDT in the United States be eliminated *within two years* (emphasis added), except those uses "essential to the preservation of human health or welfare and approved unanimously by the Secretaries of the Departments of Health, Education, and Welfare, Agriculture, and Interior." It recommended that certain other persistent pesticides—including aldrin, dieldrin, endrin, heptachlor, chlordane, benzene hexachloride, lin-

dane, and compounds containing arsenic, lead, or mercury—
be restricted "to specific essential uses which create no
known hazard to human health or to the quality of the environment" and which are unanimously approved by the
same three secretaries. It named thirteen specific pesticides
—including DDT, aldrin, dieldrin, and Mirex—as potential
carcinogens, and seven others, including 2,4,5-T, as potential teratogens. And, in contrast to the Academy report's
preference for the status quo in pesticide regulation, it recommended that approval by all three relevant secretaries—
HEW, Interior, and Agriculture—be required for all pesticide registrations, a sharp break from past practice of vest-
ing all authority in Agriculture. Ironically, the Mrak Report
ended up doing what the Environmental Studies Board had
wanted the Academy report to do: It questioned the wisdom
of subjecting the population to a long-term experiment. As
the Mrak Report expressed it: "The field of pesticide toxicology exemplifies the absurdity of a situation in which 200
million Americans are undergoing life-long exposure, yet
our knowledge of what is happening to them is at best
fragmentary and for the most part indirect and inferential.
While there is little ground for forebodings of disaster, there
is even less for complacency."[48]

Why was the Academy report so weak? Some observers
have suggested that part of the explanation may lie in a lack
of resources. In comparison with the Mrak commission, the
Academy's committee was poorly manned and poorly
financed. Whereas the Mrak commission had a staff of fifty-
three to support it, including seven full-time professionals
and eight part-time professionals, the Academy's Jensen
committee had only two part-time professionals plus secre-
tarial support. Moreover, whereas the Mrak commission and
its various panels had some forty-five members, the Jensen
committee had only fifteen. The dollar gap was about three-
to-one in favor of the Mrak commission. Final costs for the
Jensen report were about $70,000; for the Mrak commission
about $205,000, and that figure does not include the salaries
of staff members who were paid through their agency budgets rather than through the commission.[49] This disparity in
resources may partly explain why the Mrak Report was so
much more thorough and better documented than the Jensen report (though the chief reason for that difference

219

seems to be that the Mrak commission deliberately sought a well-documented report while the Jensen committee sought a concise statement). But it does not go very far toward explaining why the Mrak Report was so much tougher. Even a brief report can be explicit and hard-hitting.

The weakness of the Academy report is partly attributable to the committee's frantic desire to avoid saying anything that might offend powerful interests. Like many committees, it polarized along philosophical lines, with some members pushing the use of chemicals and some stressing ecological concerns. The result was constant compromises that led to statements so inoffensive that they said virtually nothing. The secret minutes abound with pussyfooting comments, often made by members who were satisfied with the status quo and thus wanted nothing significant said.

The desire to avoid controversy, coupled with the fact that Jensen made it clear that a minority report would be undesirable, led inevitably to a bland report full of least-common-denominator statements that almost everybody could agree to. Perhaps the extreme example of this kind of generalizing to the point of meaninglessness occurs in a section of the report devoted to "Need for Pesticides." There one finds assertions that pesticides have brought about "spectacular control of diseases" and "unprecedented" increase in agricultural productivity—"no adequate alternative for the use of pesticides for either of these purposes is expected in the foreseeable future."[50] The ordinary reader would probably not notice a peculiar subtlety in this statement—it is not talking about *persistent* pesticides, the supposed subject of the report, but about pesticides in general. And it has shifted the focus of the discussion very deliberately. At one meeting chairman Jensen himself expressed agreement with the idea that "We should cover pesticides and not the persistent ones only."[51] The result of this dodge was that the report could then say that pesticides were needed—a proposition that would wring agreement even from Rachel Carson—without really coming to grips with the more difficult problem of whether *persistent* pesticides should be phased out and to what extent.

The likelihood of a weak and nonspecific report was further increased by the committee members' perception of their roles. John E. Blodgett, a specialist on pesticides policy at the Library of Congress, conducted a careful analysis of the committee as part of a doctoral dissertation submitted to Case Western Reserve University in 1972.* He asked seven members of the Jensen committee what satisfaction they had derived from their service. Instead, they talked of a "learning experience," or cited the "prestige" of appointment to an Academy committee, or mentioned a "sense of duty." Not one said a word about trying to affect policy.

Some conservationists have contended that the committee went "soft" on pesticides because the information it received was faulty. Various counts have been put forth purporting to prove that the group of eighty-three experts interviewed by the committee was unbalanced, with a majority of those experts coming from the agricultural and industrial organizations interested in promoting pesticides. However, the disparity in numbers between "pro" and "anti" pesticide experts consulted does not seem over-whelming. A more significant factor affecting the kind of information available to the committee was probably the occasional difficulty it experienced in trying to obtain data from reluctant government agencies and industry.

The minutes of one meeting indicate that the National Agricultural Chemicals Association informed the group that "information provided federal agencies by industry in confidence could not always be made available" to the Academy[52]—an interesting revelation in view of the fact that Academy officials often assert that the opposite is the case: they say the Academy provides a valuable forum because industry is often willing to divulge secrets to the Academy that it would never dare tell a public agency. The committee also had difficulty getting objective information from government agencies since each agency tried to present its program in the most attractive light and was reluctant to

---

*We are particularly grateful for assistance rendered by Blodgett. In gathering material on the performance of numerous federal pesticide committees, Blodgett interviewed thirty-one scientists associated with four Academy committees; he graciously made his findings available to us. All responsibility for the interpretations presented in this chapter rests, of course, with us.

discuss weaknesses in its operations or data. "There is no doubt that USDA has informal data they do not want released to us," complained one committee member about monitoring results.[53] Lack of hard data on weak spots does not seem to have loomed as a major concern to most committee members, but it undoubtedly contributed to the committee's inability or reluctance to make explicit criticisms.

However, the chief reason for the committee's failure probably lies in its composition. A majority of the members came from industrial or agricultural backgrounds that would incline them to look upon pesticides as a boon to the production of food and fiber. Of the fifteen members, two— Edwin F. Alder, of Eli Lilly, and Robert P. Upchurch, of Monsanto—were directly employed by the agricultural chemical industry. A third, Louis Lykken, had worked for twenty-five years for the Shell Chemical Company and its affiliates before joining the faculty of the college of agriculture at Berkeley. A fourth, E. Paul Lichtenstein, a University of Wisconsin entomologist, received grant support from Shell Chemical Company and the Agriculture Department. And at least five others (some would say more) had backgrounds in agricultural production. This includes Jensen, who held the pivotal post of chairman. "Even though all these men were highly qualified and respected specialists," most of them were insensitive to ecological issues,[54] complained the Audubon Society's Clement. Only two members of the committee—Tony J. Peterle, of Ohio State University, and Don W. Hayne, of North Carolina State—could be considered to have an ecological orientation. Unfortunately, they were not the aggressive sort (like Cottam) who could make a strong case for their point of view. One even confessed to Blodgett that he feared he would not be able to fulfill the ecological role adequately. The other said he felt outnumbered by a pro-pesticides majority.

Two years after the noncommittal Jensen report had been published, the Academy flip-flopped again and issued a report that was sharply critical of DDT contamination of the oceans. The report—entitled "Chlorinated Hydrocarbons in the Marine Environment"—was prepared by a panel of the Academy's Committee on Oceanography (now the Ocean

Affairs Board). It warned that "as much as 25 percent of the DDT compounds produced to date may have been transferred to the sea" and that the remaining 75 percent might end up there. The amount of DDT in marine biota has already "produced a demonstrable impact" on fish, crabs, shrimp, birds, and other sea creatures, the report said, and predictions of potential hazards "may be vastly underestimated." Consequently, the report called for "a massive national effort" immediately "to effect a drastic reduction of the escape of persistent toxicants into the environment, with the ultimate aim of achieving virtual cessation in the shortest possible time."[55] The report was vague on just who was supposed to do what to curb the pollution it decried, but it nevertheless constituted the strongest warning about the hazards of pesticides yet issued by the Academy. It may even have overstated the case. Pesticide proponents protested privately to the Academy that the report failed to document its alarming allegations, and in late 1973 the Ocean Affairs Board began preparing a clarification which, according to insiders, may back down somewhat by modifying the estimated amount of DDT present in the oceans and by admitting that natural fluctuations in marine life are so great that it is difficult to measure the impact of DDT alone. "The sense of alarm in the report is no longer justified," one source has said.

Why was the report so alarmed about DDT in the oceans? The most likely explanation lies in the people who wrote the report and the purpose for which they wrote it. The panel consisted primarily of marine specialists—scientists who make no use of pesticides themselves but who would be acutely sensitive to pesticide intrusions in to the ocean.* It also contained at least one activist on the pesticide issue,

---

*The panel was chaired by Edward D. Goldberg, geochemist at Scripps Institution of Oceanography, and included: Philip Butler, marine biologist and research consultant; Paul Meier, professor of statistics and director of the biological sciences computing center at the University of Chicago; David Menzel, oceanographer at Woods Hole Oceanographic Institute; Gerald Paulik, fisheries biologist at the University of Washington; Robert Risebrough, a marine biologist at the Berkeley campus of the University of California; and Lucille F. Stickel, zoologist at Pautuxent Wildlife Research Center.

namely Robert Risebrough, who had been one of the sternest critics of the 1969 Jensen report. Moreover, the report was prepared for an environmental—rather than an agricultural—purpose. The panel participated in the month-long study of critical environmental problems (SCEP) held in 1970 by the Massachusetts Institute of Technology, and some SCEP participants contributed substantially to the panel's report. Thus the report is something of an aberration when viewed against the background of previous Academy pesticide reports. It illustrates that the Academy's stand on a given issue will much depend on what part of the Academy looks at a problem, and with what purpose in mind.

The Academy's total contribution to the national debate over pesticides which emerged during the 1960s was obstructive to efforts at reform. The Academy consistently lagged behind other eminent scientific groups in defining pesticide hazards, and its reports were frequently marred by bias or by unwillingness to take controversial stands. To be sure, at least two reports, the 1967 fire ant report and the 1971 report on chlorinated hydrocarbons in the marine environment, were sharply critical of pesticide contamination. In fact, both were produced by committees that were "loaded" with scientists who were apt to oppose the pesticide use under examination. But the main thrust of the Academy's advice on pesticides has reinforced the status quo. The committee whose members attacked Rachel Carson in 1962–1963; the committee which backed the Agriculture Department's dubious handling of pesticide regulation in 1965; and the committee which refused to help USDA devise a strategy for phasing out persistent pesticides in 1969 all threw their prestige behind "business as usual" and opposed the reformers. Moreover, the Academy's acquiescence in suppression of its fire ant report deprived reformers of a tool that might have helped overturn an eradication campaign that is widely condemned in the scientific community.*

------

*Academy officials say they initiated a new publication policy in the 1970s: They retain the right to publish any report, except those that are classified,

The chief reason for the Academy's obstructive posture is that its most significant pesticide projects have been housed within the division of biology and agriculture, the section of the Academy that has the closest ties to agriculture and industry—the very interests which were largely responsible for creating the pesticide problem in the first place. This situation appears to have been improved by putting the latest pesticide study under the Environmental Studies Board. The study, which was launched in 1971 and may be completed in 1975, will be a comprehensive cost benefit analysis that will consider the scientific, social, and economic implications of pest control on a broader scale than any previous Academy study. One topic of study will be alternatives to the use of chemical pesticides. Experts are being drawn from a variety of disciplines—public health, ecology, agriculture, forestry, and economics, among others.

But President Handler continues to throw his weight against the environmentalists. In an interview with *Nation's Business,* he called the stress on pesticides "exaggerated" and suggested that the issue was simply a "handful of bird species that seem to be endangered."[56] And in a major address to the American Association for the Advancement of Science he said:

"I share the concern for the possible hazards of DDT—but not the hysteria of those who demand an absolute prohibition against its use before an acceptable substitute is available. The predicted death or blinding by parathion [a toxic substitute for DDT] of dozens of Americans last summer must rest on the consciences of every car owner whose bumper sticker urged a total ban on DDT."[57]

The trouble with that formulation of the issue—which parrots the line that the chemical industry has adopted—is that it implies, wrongly, that the anti-DDT campaign is being led by the ill-informed little guy with a bumper sticker who has succumbed to hysteria. This conveniently ignores the fact that three distinguished scientific panels—the Mrak commission, the SCEP group, and a special DDT panel

---

no matter what the funding agency wants done with the report. If the Academy adheres to this policy, similar problems may be avoided in the future.

convened by EPA—have all recently called for a drastic reduction in use of DDT.* Handler's emphasis on the difficulties of reform looks like more of the same obstructionism from an institution that has generally ignored the excesses of the pesticide polluters.

———————

*The Mrak commission called for elimination "within two years" of "all uses of DDT and DDD in the United States excepting those uses essential to the preservation of human health or welfare and approved unanimously by the Secretaries of the Departments of Health, Education, and Welfare, Agriculture, and Interior" (Recommendation 3); the SCEP report said, "We recommend a drastic reduction in the use of DDT as soon as possible...." (*Man's Impact on the Global Environment*, Report of the Study of Critical Environmental Problems, MIT Press, 1970, p. 25); and the EPA panel concluded that, while DDT does not pose an imminent hazard to human health, it does pose "an imminent threat to human welfare in terms of maintaining healthy, desirable flora and fauna in man's environment. The panel urged that DDT use be rapidly curtailed "with the goal of virtual elimination" (see *Science*, December 10, 1971, p. 1109).

CHAPTER TEN

Airborne Lead:
A Failure to Delineate the Hazards

"The National Academy lead study is a dramatic example of how our most prestigious scientific body is incapable of taking a stand regarding the risks associated with introduction into the environment of substances that damage people in insidious epidemiological ways."

> —Paul P. Craig, chairman of the Environmental Defense Fund's Committee on the Environmental Impact of the Automobile, in an article in *Saturday Review*, October 2, 1971.

"We knew—albeit belatedly—that the report on lead in the environment did have at least one serious defect."

> —Philip Handler, president of the Academy, in his *Letter to Members*, a private communication, May 1972.

On September 7, 1971, the National Academy of Sciences issued a massive report—entitled "Airborne Lead in Perspective"—that was intended to help the Environmental Protection Agency (EPA) decide what to do about controlling lead pollution.[1] The report immediately touched off a controversy among specialists concerned with atmospheric lead and its effects on human health. Lead industries touted the report as evidence that concern over pollution from their products was misplaced, while environmentalists charged that the report was a whitewash engineered by industrial scientists who wrote crucial portions of the text.

The lead report was the first in a series of surveys and evaluations of specific pollutants prepared by the Academy under contract to the Environmental Protection Agency. The overall responsibility for the project—which by 1973 was surveying some sixteen pollutants—was vested in a Committee on the Biologic Effects of Atmospheric Pollutants (BEAP), housed in the Academy's Division of Medical Sciences. But the actual drafting of each report was to be done by a special ad hoc panel set up to deal with the specific pollutant involved.

The problems of the lead report can be traced, in part, to the original appointment of panel members. Critics have questioned both the competence of the panel and its objectivity. The names of prospective panelists were generated primarily by staff members of the Division of Medical Sciences through what is known colloquially around the Academy as the "old boy" network. Staff members solicited suggestions from several sources—fellow staffers, members of the parent BEAP committee, and other scientists whose judgment they trust—and the list of names thus compiled was then narrowed down to a handful that were asked to serve by the Academy.

There is some feeling that the "old boy" network didn't cast its net far enough in this case. Strangely enough, the panel included not a single person who had previously done much work on the specific problem to be addressed: *airborne* lead. The chairman of the panel was Paul B. Hammond, a veterinary pharmacologist at the University of Minnesota College of Veterinary Medicine, who had studied lead poisoning in animals. "My concern with airborne lead as such had been minimal," Hammond says. "There's hardly anybody on the committee, as I recall, that really has great interest or competence in this field [of airborne lead pollution]," complains Robert L. Metcalf, chairman of a team of scientists who conducted a review of the report for the Academy before it was released. "It certainly should have involved more people who are actually measuring lead and looking at the hazards of it in the environment."*

The panel was also remarkable for its exclusion of scientists who sounded the alarm about alleged dangers of atmospheric lead concentrations. A partial list of such scientists would include John Goldsmith, head of the California Health Department's epidemiology unit; Henry A. Schroeder, head of Dartmouth College's Trace Metal Laboratory; Clair Patterson, geochemist at the California Institute of Technology; Paul P. Craig, a physicist who heads the Environmental Defense Fund's lead committee; and T. J. Chow, a geochemist at Scripps Institution of Oceanography. Some of these men were never seriously considered for membership on the panel because they lacked the necessary biological expertise to contribute much to a study focusing on biological effects; others were considered but rejected for reasons that have become murky as various Academy officials have

---

*Other members included Arthur L. Aronson, a veterinary pharmacologist at Cornell University College of Veterinary Medicine, who had been a former graduate student of Hammond's and had studied the treatment of lead poisoning in animals with chemical agents; J. Julian Chisolm, Jr., a Johns Hopkins pediatrician who is a leading authority on biochemical effects of lead poisoning, with particular emphasis on the problem of children who eat paint, plaster, and other substances containing lead (a form of behavior known as pica); John L. Falk, a physiological psychologist at Rutgers University who had had no previous experience with lead; Robert G. Keenan, an analytical chemist at an industrial consulting firm, George D. Clayton & Associates, Incorporated, of Southfield, Michigan, who is a specialist on measuring and monitoring techniques; and Harold H. Sandstead, a Vanderbilt University internist and nutritionist who had studied lead poisoning from moonshine whiskey.

given out conflicting explanations in the wake of controversy over the lead report. Nevertheless, it seems likely that Goldsmith and Schroeder were excluded at least partly because staff members of the Division of Medical Sciences thought they might prove too "disruptive" in pressing their points of view. That, at least, is what one Academy staffer told *Science* magazine, and Hammond, the panel chairman, confirmed that "The word I got was that they were unacceptable."[2] As Hammond told us: "I don't presume to know why they do these things. I think Goldsmith would have had something to contribute." On the other side of the ledger, the Academy also excluded one of the leading scientific spokesmen for the lead industries, Robert A. Kehoe, a longtime medical consultant to the Ethyl Corporation, which is a major producer of lead additives for gasoline.

The panel held its first meeting in July 1970 and decided that it would bolster its expertise by soliciting help from consultants. Ultimately some thirty-five consultants were enlisted. Some were merely asked to comment on early drafts of the report; some contributed specialized information; and at least nine actually wrote the initial drafts of various sections of the report. The balance of viewpoints among these nine writing contributors was not as finely drawn as the balance on the lead panel itself. The list included at least one outspoken critic of lead pollution—Chow—but he was more than offset by four scientists who were employed at the time by DuPont or Ethyl, the two major producers of lead additives for gasoline.*

One of the industrial scientists, Gordon J. Stopps of DuPont, actually wrote crucial parts of the report. Stopps seems to have stumbled into this role somewhat by accident. He was a member of the parent BEAP committee and in this capacity had been assigned to serve as an associate editor for the project, a role in which he would have monitored the panel's activities and commented on drafts. But then, as

---

*The four included Gary Ter Haar, of Ethyl, who contributed part of a chapter on lead in the ecosystem; John M. Pierrard, of DuPont, who contributed part of a chapter on nonbiologic effects of lead; Kamran Habibi, of DuPont, who contributed an appendix on measurement of the size of airborne particles; and Gordon J. Stopps, of DuPont, who ended up writing two key sections of the report.

luck would have it, a Harvard researcher who had agreed to serve on the panel backed out of the project, and Hammond, in a hurry to get the report completed, asked Stopps to take over writing a section on adult epidemiology—a critical assignment, since it deals with much of the evidence concerning whether there are human health problems associated with lead. Subsequently, the panel also asked Stopps to draft a chapter on the air pollution role of lead alkyls (the additives themselves, as distinct from their combustion products). "It was never intended that he [Stopps] be a writer in any regard," Hammond says, but "as long as we had him involved in writing the epidemiology, we said, what the hell—he might just as well write the lead alkyl part too, because in our conversations around the table it became apparent that he knew more about this than anybody else did." In contrast to the other consultants who drafted parts of the report at long range, Stopps became a de facto working member of the panel. He participated in all panel activities, including the framing of conclusions. He had probably done more work on the subject of airborne lead than any of the original panel members.

The key role assigned to Stopps set off a flurry of debate in Academy circles. Harriet Hardy, a prominent expert on metal poisoning who had been asked to review early drafts of the lead report, complained to the Academy that the list of authors was "top heavy" with industry scientists, particularly Stopps.[3] Chow wrote a letter complaining about "possible conflict of interests" involved in letting a DuPont employee "play such a key role in preparing this report."[4] And at least one member of the parent BEAP committee told us he protested the inclusion of Stopps. But the Academy would not back down. Louise H. Marshall, the staff officer for the lead panel, told Chow: "It is presumptuous to assume that the Panel as a whole would be dominated by the opinion of any one of its members. Rosters of committees and panels consist of people with high competence in specific fields regardless of where they work and the appointment is made with the understanding that the person is thought to serve as an individual and not as a representative of his organization, whether it be a university, public agency, or industrial organization. The important thing is that there be no hiding of any possible source of bias."[5] Stopps, who subsequently left DuPont to take a public health job in Canada, also pointed out that he was the only industrial scientist actively involved in the panel's discussions. "It's highly unlikely that one person could railroad the whole panel," he said.[6] But critics of the lead panel remain skeptical. "How could he [Stopps] be neutral," Hardy told one interviewer. "He has written and written for years that there's nothing harmful about tetraethyl lead.... It's just not possible for him to act purely as a scientist."[7]

The panel at first had difficulty focusing on atmospheric lead, the main problem it was supposed to consider, probably because it had so little expertise on that issue. Instead, it got sidetracked on such problems as lead poisoning from paint and moonshine whiskey, the pet subjects of various panelists. One atmospheric scientist who read an early draft of the report called it "pitiful.... The first outline didn't even treat in depth atmospheric lead as a problem.... Some of the guys didn't even know that a large share of the particles in urban atmospheres are from lead." Dunham agrees that the early meetings of the panel "got almost preoccupied with the problem of pica—my own impression is that it threatened to overweight the whole report." But he feels that the panel ultimately worked its way back to a balanced treatment of the problem. The report was issued under the title "Airborne Lead *in Perspective*" (emphasis added) to indicate that it devotes much attention to other topics. The panel explained—or perhaps rationalized—that one could only understand the subtle effects of atmospheric lead concentrations by first considering the more acutely harmful effects of lead poisoning from other sources.[8]

The drafts submitted by the various contributors were subjected to close scrutiny that was intended to insure accuracy and eliminate any bias. The drafts were read and revised line-by-line by the entire lead panel and were circulated for comment to outside experts, members of the parent BEAP committee, EPA scientists, other divisions of the NRC, and the Report Review Committee, the Academy's top reviewing board. Two anonymous outside reviewers—Hardy and Kehoe, who represent opposite opinions on lead—criticized the report as it progressed through various drafts. Hardy says she battled repeatedly over Stopps's epidemiology section, since she felt he was emphasizing relatively old studies that tended to exonerate lead as a hazard while ignoring or downplaying more recent studies which suggested—to her at least—that lead is a potential menace. There was considerable give and take in the whole process, and some chapters were substantially rewritten. Thus the report was, in large measure, a group effort. But the initial authors of various sections were unquestionably in a powerful position to set the tone and thrust of the report.

The document which finally emerged on September 7, 1971, was rambling, repetitious, and written in a style which obscures rather than illuminates critical issues. The Academy was under pressure from EPA to get the job completed, so the report was first issued as an "advance copy," containing 333 typewritten pages plus lengthy appendices. A 330-page printed version (including appendices), which differed in only minor respects from the advance copy, was published in early 1972. The report lists six hundred references.

The document is difficult to summarize. Some have likened it to the Bible, in that it is less a position paper than a composite of views containing data that can be used to support a variety of stances. But most observers thought they detected a tone of complacency toward lead hazards. To begin with, the report's summary chapter concludes that "The average lead content of the air over most major cities apparently has not changed greatly over the last 15 years. . . . We are, in short, not dealing with a rapidly shifting scene in this respect."[9] This conclusion was disputable because, as we shall see, the most recent data suggests—though it does not conclusively prove—that the opposite might be the case, namely, that atmospheric lead concentrations in urban areas may be rising sharply. The report then went on to imply that lead from gasoline additives is a relatively small problem compared with other forms of lead pollution and that reduction of such emissions might therefore not be of high priority. As the report's summary chapter expressed it: "Two to three times as much lead is added to the total environment in the form of paint pigments and metallic products as in the form of lead alkyls. . . . Any proposal for the removal of lead from automotive fuels to rid the environment of lead pollution must take into consideration the fact that the fate of other lead products, such as paints and manufactured items, is largely unknown."[10] This, too,

was a controversial statement. Chow, for example, complains that, while the statement is true, it is highly misleading, since lead that is inhaled stays in the body more readily
than lead that is ingested from other sources, such as paint
pigments.[11]

After thus establishing that airborne lead is neither a
particularly large nor a rapidly worsening form of pollution,
the report's summary chapter examines the effects of lead
on man. Surprisingly, the version of the report released in
September 1971 contains no explicit statement concerning
whether airborne lead poses a threat to the general population. Sentences here and there imply that the general population is in no danger, but at no point does the report
directly pose and answer this question. The preface does
say, by way of explaining why the panel spent so much time
on nonairborne sources of lead, that "lead attributable to
emission and dispersion into general ambient air has no
known harmful effects." But the report itself does not contain any statement to that effect. This deficiency was readily perceived by writers in the Academy's Office of Information who found themselves unable to compose a sensible
press release describing what, if any, hazards the panel had
discovered. The writers therefore went back to Hammond,
the panel chairman, and Marshall, the staff officer, for further explanation—with the curious result that the press
release, which was approved by Hammond and Marshall,
contains a much more explicit statement than can be found
in the advance copy of the report. "The high concentration
of lead in the air of central cities ... poses no identifiable
current threat to the general population,"* the release says.
"The average American, even in the cities, consumes more
lead in food and beverages than he inhales from the air....
For an average city resident, the total exposure produces a
blood lead concentration about half that necessary to cause
biochemical changes in the body and one-fourth the level at
which symptoms of lead poisoning begin to occur."[12] As we
shall see, this, too, was a controversial assessment. Other

---

*This phraseology was picked up in the final printed version of the report,
which has a clause inserted stating that "Although the concentration of
lead in the air of cities poses no identifiable current threat to the general
population ..." (p. 209).

experts, some with credentials as good or better than those of the Academy panelists, look at precisely the same evidence and conclude that, while people are not yet dropping in the streets, the margin of safety for the general population is disturbingly small.

The Academy report did find that "two special categories of people are exposed to lead of atmospheric origin to a degree that seems undesirable."[13] The first category included garage workers, traffic policemen and others who are exposed continuously on the job to high concentrations of airborne lead, a circumstance which may result in blood lead concentrations sufficient to cause biochemical changes in the body. But the report softpedaled this conclusion by adding that: "Even within this small population, this level of blood lead concentration is probably attained by only a relatively small proportion of those exposed. To reach the very high blood lead concentration compatible with clinical lead poisoning would probably require . . . a fivefold increase beyond current levels in lead assimilation for a long period."[14]

The other special category for which airborne lead poses "a significant threat" consists of infants and young children in the inner cities. The report noted that recent surveys of children in large cities indicate that many have blood lead concentrations in the range associated with biochemical changes (though not with symptoms of lead poisoning). It suggested that these children may have swallowed soil or street dust contaminated by airborne lead, may have inhaled the lead directly from the air, or may have eaten lead pigment paint chips. Although ingestion of lead pigment paints is clearly the main cause of acute lead poisoning in young children, the report estimated that daily ingestion of highly contaminated dust over a long period could produce symptoms of lead poisoning even without any other sources of exposure. Unfortunately, the amount of dust or soil eaten or inhaled by young children was "totally unknown."[15]

The Academy's document was hardly a rousing call for action to reduce atmospheric lead concentrations. Initial press reaction generally interpreted the report as indicating lead was no hazard. The Associated Press, the nation's largest wire service for newspapers and broadcasters, described the report as "minimizing pollution from leaded gasoline."[16] *Chemical and Engineering News* headlined its story: "Lead in Air No Threat."[17] The *Sioux Falls* (S.D.) *Argus-Leader* editorialized: "The National Research Council said the worry about leaded gasoline was rubbish."[18] And leading financial papers felt lead had been exonerated. "Lead in Atmosphere an Overstated Peril, Study Group Asserts," was how *The Wall Street Journal* headlined its story.[19] And *Barron's*, a financial weekly, editorialized that the report "went a long way toward exploding the alarmist myths" by demonstrating that "The perils of tetraethyl lead ... are illusory."[20]

Industry executives were openly jubilant. The Ethyl Corporation issued a press release asserting that the report "confirms the position which Ethyl has taken throughout the lead-in-gasoline controversy, that the use of lead anti-knocks does not 'endanger the public health or welfare.'"[21] The stock market reacted as if Ethyl had been exonerated. Ethyl common stock jumped from 20⅝ on the last trading day before the Academy report was released to 24⅝ the day after the report was released. And the Lead Industries Association, Incorporated, soon launched a traveling press conference to debunk the idea that lead in gasoline is a hazard, citing the Academy report as prime evidence.[22]

Environmentalists and several prominent lead scientists were understandably dismayed. Chow complained that the Academy's press release conveyed "a false sense of security and well-being" and that the report's summary chapter glossed over the dangers of airborne lead.[23] Goldsmith complained that "major fundamental questions about lead are blurred by the massiveness of the report, and are further blurred by the press release."[24] He also called the report "almost unbelievably selective—and therefore biased."[25]

However, some of the sharpest criticism came from the Academy's own Report Review Committee (RRC) in internal critiques submitted before the report was published. The RRC, which is the Academy's highest reviewing body, appointed a panel of five Academicians to review the lead report. The panel was headed by Metcalf, a University of Illinois zoologist who can claim some expertise in pollution matters himself. Metcalf was one of the organizers of a large interdisciplinary project at the University of Illinois—perhaps the largest of its kind in the country—which had received some $1.2 million from the National Science Foundation for a three-year study of the environmental impact of lead. Metcalf recalls that his review panel was "quite aware that it [the lead document] was going to be a poor report." The reviewers had two major complaints, according to Metcalf. The first was that the report was "almost noncommittal" about explaining the significance of the mass of data it surveyed. Metcalf calls it "an enormous technical accomplishment to review all that literature," but he complains that the report "failed miserably to form any sort of a precise conclusion." "I have a feeling that in the report there are all the danger signs that point that somewhere there is a threshold beyond which lead pollution just cannot go without widespread harm and yet the total conclusions don't reflect that at all," he says. "What does all this mean? There's no point in just being a high-priced data collector."

The second major objection involved the report's assertion that air lead levels in urban areas have not changed greatly over the past fifteen years. In submitting the RRC's original review of the report, Metcalf said that this conclusion wasn't in line with data quoted by Chow from San Diego and with data collected elsewhere by the Public Health Service. Later, after the review process had been completed, Metcalf received even more disturbing evidence of rising lead levels from the so-called Seven–City Study, a joint government–industry project which is one of the most extensive surveys of lead in the air and in human populations ever conducted. Preliminary data from the study were first released at a public hearing in Los Angeles on November 9, 1970, midway through the Academy panel's deliberations, but these data do not seem to have come to the panel's attention while it was preparing its initial drafts. The preliminary findings were later given wide circulation when EPA distributed a press release about the study on June 4, 1971. This was after the Academy panel had concluded its meetings but well before its report was released. The data revealed that airborne lead increased at seventeen of nineteen measuring stations in Cincinnati, Los Angeles, and Philadelphia when 1968–1969 measurements were compared with measurements made in 1961–1962. In Los Angeles, increases ranged from 33 to 64 percent; in Philadelphia, from 2 to 36 percent; and in Cincinnati, from 13 to 33 percent.[26] The increases were substantial and did not appear to jibe with the Academy panel's conclusion that air lead levels were holding even.

When Metcalf saw these findings, he photocopied the data and sent the copy off to the Academy; he also called the Academy and tried to get the lead report held up until the new data could be looked at—all to no avail. "We just couldn't seem to break down this ponderous machine that had been set up," he recalls. "The report had to go right out whether it was wrong or right. This disturbed me a great deal."

The report did include two grudging, last-minute references to the air lead data from the Seven–City Study. At one point, it said that air lead concentrations "at some individual sites" were higher in 1968–1969 than in 1961–1962—a rather slighting way to take note of the fact that readings were higher at seventeen of the nineteen sites.[27] And in the summary chapter, it said that, while we are not dealing with "rapidly shifting" air lead levels, "more recent information [i.e., the Seven–City data] is not in complete agreement with this conclusion and might slightly modify it."[28]*

Just why the panel at first ignored, and then downplayed, the Seven–City air lead data is a mystery. According to conflicting explanations offered by various principals, the air lead information was ignored either because the authors of the study refused to make it available while the committee was deliberating; or because they offered it but the Academy panel neglected it; or EPA, which had access to the data, failed to make it available to the Academy; or because EPA sent it to the Academy but the Academy lost it.

The apparent complacency of the Academy report contrasted sharply with the conclusions reached by another eminent scientific group. Three months after the Academy

---

*The final version beefed up its references to the Seven–City Study slightly. The statement about "some individual sites" showing higher readings was replaced with a more precise statement giving the actual numerical findings of the study. But the final version stressed in a footnote that the data is "preliminary" and it retained the assertion that "We are . . . not dealing with a rapidly shifting scene."[29]

report had been issued, the Illinois Institute of Environmental Quality, a state agency, issued a 149-page report entitled "A Study of Environmental Pollution by Lead." The report had been prepared by an interdisciplinary group at the University of Illinois—the same group that was conducting the large-scale study of lead pollution under NSF sponsorship. Although much of the basic data was the same in both reports, the Illinois group came up with a more alarming interpretation of its significance. Whereas the Academy panel found "no identifiable current threat" to the general population, the Illinois group warned that "The margin of safety appears to be low" for the average citizen. Whereas the Academy panel had stressed that paint and metallic products add more lead to the environment than gasoline additives, the Illinois report said that lead from gasoline is "a particularly mobile form and is therefore more significant in the environment than solder used on tin cans, ammunition, and pigments all of which weather and become mobile in the environment only very slowly." And, whereas the Academy report had claimed that air lead levels have not changed greatly over the past fifteen years, the Illinois report explicitly rejected this conclusion as "not warranted." It claimed that an increase in traffic density and in leaded gasoline consumption made it virtually "inevitable" that atmospheric concentrations have increased; it cited the Seven–City data and earlier studies as evidence that lead is increasing in the atmosphere; and it cited an increase in lead found in snow samples, soils, and tree rings as "indirect evidence" that air lead levels have increased. The Illinois report's overall conclusion was that "people are being subjected to lead concentrations, particularly in food and urban air, which are not far below those levels which result in overt symptoms of plumbism [lead poisoning]." Since some 90 to 95 percent of all airborne lead originates from automobile exhaust emissions, the Illinois report recommended that "steps should be taken to drastically reduce the use of leaded gasoline gradually over the next decade."[30]

The Academy's apparent complacency also contrasted with concern shown by other qualified groups and individuals. Julian M. Sturtevant, professor of chemistry, molecular biophysics, and biochemistry at Yale University, said he was "amazed" that the Academy panel found no "identifiable current threat" to the general population despite the fact that the panel indicated that the average city dweller already has blood lead concentrations half that necessary to cause biochemical changes and one-fourth that at which actual symptoms of lead poisoning occur. "It seems to me that the remarkable conclusion reached allows no margin of safety for upward deviations from the average, or for the well-known wide variations in susceptibility to toxic materials shown by individuals in any population," he contended.[31] Similarly, the Environmental Defense Fund contended that the safety margin was "extremely narrow, and statements that current levels are safe are without foundation."[32] And Nobelist Joshua Lederberg once agrued that "The proven safety margin is so small that if the procedures routinely applied to new drugs by the Food and Drug Administration were applied to lead tetraethyl, it could hardly be approved unless it carried enormous lifesaving benefits in compensation."[33]

Ironically, although almost all public comment on the Academy report viewed it as exonerating lead, Hammond, the panel chairman, believes that "Based on the information we provided there is a clear basis for doing something about lead in the air." He cites data in the report which indicate, he says, that lead levels in the air, soil, and street dust of many cities are already sufficiently high to be dangerous. "So I don't think it's a question of holding the level where it is," Hammond says. "I think that the present level has been shown, in our document, to constitute a hazard. In other words, there are seventy-seven American cities that have been reported to have street dust and soil concentrations of lead somewhere in the neighborhood of 2000 parts per million [ppm]. And we constructed a rationale to show that if little kids eat this stuff in milligram-per-day quantities it could have a substantial effect on their body burden of lead. . . . A fraction of these little kids in the inner city already are carrying around more doggone lead than you would want them to carry. . . . Now we don't know how much street dust kids eat or how much they inhale—nobody's done these studies—but I'm saying that there's a basis for concern here."

At least two critics of lead pollution agree with Hammond that the Academy report reveals the dangers of lead. Harriet Hardy told us that, "If you take time to read the report, everything is there that you'd like to see. You just have to have enough knowledge of the field to interpret it." And Derek Bryce-Smith, a leading British opponent of leaded gas, believes the report's analysis of the hazards faced by adult workers and young children in the inner cities, among other conclusions, "would seem to me to make an overwhelming case for immediate abolition of all lead additives from petrol." In fact, Bryce-Smith has used the Academy report to buttress his campaign against lead in the British Isles.[34]

Still, the more common interpretation is that the Academy report exonerated lead, and this raises the question of why the Academy panel "failed miserably" (in Metcalf's terminology) to make it clear that airborne lead is a hazard that should be curbed. There is probably no single explanation; instead, a number of factors interacted to enhance the likelihood of an indecisive report. One possible explanation is that the panel had too little expertise on airborne lead to feel confident in—or perhaps even interested in—reaching firm conclusions. Another is that Stopps, simply by virtue of superior knowledge of airborne lead, exerted undue influence on the tone and thrust of the report (though opinions are mixed on this point). A third is that key portions of the text were written by industrial scientists, including Stopps, who were predisposed to go easy on lead. A fourth is that the illness of Hammond's wife, who died during the time the project was being conducted, distracted him from his duties. A fifth is that the panel, perhaps because it contained no scientists who are leading the campaign against lead, was not very aggressive in ferreting out information, as was evidenced by its failure to obtain the Seven–City data whose existence was known to industry, the EPA, and public officials in California (notably Goldsmith, who had been barred from the panel). And a sixth is that the drive for consensus which affects almost all Academy committees made it difficult to agree upon any kind of boat-rocking conclusions.

A seventh explanation—proffered by Hammond himself—is that the panel lacked the literary skill to express its concerns clearly, while the Academy staff did little to help clarify what the scientists were trying to say. Hammond calls the lack of "capable editorial writers" a "staff deficiency." The literary problem was undoubtedly compounded by the fact that the panel had to rush its work in the final stages to get the report out on time.

The likelihood of a noncommittal report was further enhanced by an eighth factor—the panel's attitude toward its task. The panel seems to have assumed that its job was primarily to review a mass of literature, while leaving it to EPA to infer the obvious conclusions from the data. "The charge to us from the EPA was to develop *background* documents [emphasis added] from which these people could extract meaningful criteria for air standards ... or any other relevant controls," Hammond says. "We were never asked to make a specific recommendation relative to what level of lead in air should constitute an acceptable maximum."

It's undoubtedly true that a panel of six scientists is not the proper group to devise air standards that must necessarily be based on economic and social factors as well as purely scientific data. But the panel chose to interpret its mandate too narrowly, for the contract covering the lead study appears flexible enough to allow expressions of concern. Even without recommending standards of its own, the panel could surely have defined the health risks associated with various levels of lead pollution, leaving it to the responsible government officials to decide whether those health risks justified regulatory action. Moreover, the panel could surely have said it considered existing air lead levels in many cities to be dangerously high (assuming that the panel agrees with its chairman on that point). Why did the panel shy away from such statements? "We're not supposed to express concern—that's not our job," Hammond says. "I, at least, conceived of this as a report to EPA.... I thought of myself as talking to guys like Horton and Romanovsky [EPA scientists] who know this business.... We were providing them with the meat from which they could extract their recommendations."

The trouble with that formulation of the Academy's role was that it deprived EPA, and the public, of specific guidance from the Academy as to the degree of health hazard posed by airborne lead. It was always possible that EPA and the Academy might differ in assessing the significance of the evidence. EPA scientists might simply interpret the data differently than the Academy panel. Or EPA, in its zeal to protect the environment, might tend to see hazards where the Academy found none. Or EPA, faced with powerful opposition from the lead and oil industries, might hesitate to launch a crackdown on lead, even though evidence in the Academy report suggested such a crackdown would be desirable. For all these reasons, the Academy, which is supposedly immune from many of the pressures that afflict EPA, should have issued a report which defined, as explicitly as possible, the health hazards of lead. If the nation's most prestigious scientific organization thinks a situation is hazardous, it should say so and not depend on a regulatory agency to infer that the situation needs correcting.

As it turned out, EPA initially proposed action that Hammond, at least, felt was inadequate. On February 23, 1972, EPA proposed regulations that would reduce the lead content of gasoline in phased steps between 1974 and 1977. The goal, according to EPA, was to reduce lead emissions from motor vehicles by 60 to 65 percent so as to bring airborne lead levels down below 2 micrograms per cubic meter—a level which EPA described as "fully protective of public health" based on "present scientific evidence."[35] (Airborne lead levels in many urban areas currently range from 2 to more than 5 micrograms per cubic meter.) As justification for its regulations, EPA produced background documents which cited many passages from the Academy report as evidence that there is a health hazard associated with high lead levels.

EPA scientists who helped draw up these background documents were privately boastful that they had managed to extract enough material from the supposedly weak Academy report to justify stiff regulation action. But Hammond, ironically, contended that EPA's action was too weak to remedy the situation described in his report. "I had thought about trying to contact Nader over my concerns about what EPA is proposing to do standardwise on this matter of accepting 2 micrograms per cubic meter as a reasonable level," Hammond told us. "If you look at our report ... I don't think that really makes a hell of a lot of sense."

In three letters to EPA, Hammond argued that the 2 ug/-m³ figure was "too high" because it posed "hazards for special groups," including traffic policemen and parking lot attendants who work in locations where the concentration of lead is apt to be high and young children who live in areas where street dust is contaminated by lead. "I am bothered by the 2 ug/m³ figure because the NAS report strongly suggests that this very level is hazardous," Hammond wrote. "...the general ambient air lead values in the 77 midwestern cities from which street dust was collected did not generally exceed 2 ug/m³ and were found to contain 2,000 ppm lead. To hold ambient air lead at 2 ug/m³ is to allow the persistence of the very conditions which the NAS lead panel judged to be undesirable."[36]

Similar doubts about the proposed level were expressed by the Illinois Institute for Environmental Quality, whose earlier report had expressed much greater concern about lead pollution than had the Academy document. The Illinois group recommended an ambient air quality standard of 1.5 ug/m³.[37]

But EPA, faced with strong industry attacks on the scientific justification for even its 2 ug/m³ target, waffled on the issue. It reproposed the regulations in somewhat modified form on January 10, 1973, then issued final regulations in yet another form on December 6, 1973. The final regulations no longer specified an ambient air lead goal of 2 ug/m³, largely because EPA found it difficult to establish just what level was needed to protect health. But the final regulations still called for a 60 to 65 percent reduction in lead usage.[38] According to Kenneth Bridbord, medical officer for EPA's National Environmental Research Laboratory, this reduction is expected to achieve essentially the same air lead level as the earlier regulations, namely 2 ug/m³.

Thus EPA was preparing to put into effect a phased reduction in the lead content of gasoline that was expected to result in an air quality level—2 ug/m³—which many competent scientists, including the chairman of the Academy's lead panel, consider dangerous. Whether the full lead panel would agree with Hammond is not known. But, if Hammond is correct in his assessment that the action proposed by EPA is not sufficient to protect the public health, surely part of the blame for EPA's failure can be traced to the Academy lead panel itself, which failed so conspicuously to produce a lucid assessment of airborne lead hazards.

CHAPTER ELEVEN

The Academy
and the
Public Interest

"Certainly no entity in American society is so uniquely placed, or has an equivalent opportunity to be of service to the nation."

—Philip Handler, in his *Letter to Members*, February 1972.

In examining the Academy, the vantage point we chose was that of the citizen who would like the Academy to bring the nation's best scientific talents to bear on societal problems and then enunciate, unflinchingly and unequivocally, the nearest possible approximation to the truth. From that perspective, it seems clear that the Academy, all too often, turns in a flawed performance. If the public interest would be served by far-sighted, highly competent, scrupulously objective studies—which, after all, is supposedly what the Academy seeks to perform—then the Academy, in many cases, falls short of serving the public interest.

The Academy has issued some reports of outstanding quality. A 1963 report on The Growth of World Population, which called for research to control rising birthrates, is credited with pushing the federal government to boost its support of birth control efforts at a time when it was politically difficult even to discuss the touchy issue. An Academy review of the efficacy of some 3,000 prescription drugs, conducted during the late 1960s, was described by Food and Drug Commissioner Charles C. Edwards as "uniquely extensive in time and scope. It is fair to say that the study lays the groundwork for some of the most significant medical reforms in many years." A 1972 report on the biological effects of ionizing radiation won praise from all sides for its soundness and fairness. And a 1973 report on motor vehicle emissions performed a major public service by warning that the American automobile industry might be adopting the most disadvantageous technology for controlling pollution.

But such documents are limited in number. We found, surprisingly, that the Academy's advisory reports often fall short of the very high quality one would expect from the nation's preeminent scientific organization; many, in fact, are mediocre or flawed by bias or subservience to the funding agencies.

The Academy has seldom, if ever, issued the kind-of seminally significant document which revolutionizes a nation's perception of public issues. There seems to be nothing in the Academy's vast output quite comparable to the 1910 report by Abraham Flexner, which led to a complete overhaul of medical education; or the 1945 report by Vannevar Bush and his colleagues, *Science, the Endless Frontier*, which led to creation of the National Science Foundation and increased support for basic research; or the 1966 report on *Equality of Educational Opportunity* by sociologist James Coleman, which provided intellectual thrust to the nation's school desegregation efforts.

Even President Handler has acknowledged that, while the Academy has had a "pretty good" record in terms of assisting the federal agencies, it has had "very little impact indeed" on such important problems as crime, the drug culture, and worsening urban and rural conditions, which caught the Academy (and the government as well) more or less napping.

Some influential Academicians believe the number of defective reports has been distressingly high. Harvey Brooks, dean of engineering and applied physics at Harvard, who has long played a prominent role on key Academy committees, once confided in a 1969 letter to Nobel Laureate George Wald: "I have a huge file of correspondence with respect to the SST sonic boom report, the defoliation report, the civil defense report, the mineral sciences report, and *hundreds of other NRC reports* [emphasis added] which I had been collecting as a 'Chamber of Horrors.'"[1]

Thus, an important lesson to emerge from this book is that one should be cautious about accepting the Academy's pronouncements as the Unchallengeable Word from On High. Academy studies run the gamut—from significant to trivial; forward-looking to old hat; courageous to timid; objective to biased. Each must be judged on its individual merit rather than accepted on faith because of the aura of prestige which surrounds the Academy.

The Academy can be faulted not just for sins of commission (the writing of defective reports) but for sins of omission as well. The Academy has seldom taken the lead in pushing for solutions to major societal problems; instead, it has often allowed itself to be used as a shield by those intent upon preserving business-as-usual.

There is abundant evidence in our case history material to support this thesis. In our examination of pesticides, for example, it was Rachel Carson who first sounded the alarm about deleterious side effects, whereas the Academy pooh-poohed the alleged danger. In the area of defoliation, it was the American Association for the Advancement of Science which led the investigation of harmful effects, while the Academy allowed itself to be used by the military and its supporters in an effort to head off the AAAS thrust. In the case of the supersonic transport, it was citizens' groups in Oklahoma and elsewhere that expressed resistance to the sonic boom, and the Academy was rushed in by the government to quiet the opposition. Then, when a scientist working on an Academy committee did raise disturbing questions about the SST's possible impact on the incidence of skin cancer, the Academy practically disowned him. And in the case of the Atomic Energy Commission's plan to bury radioactive wastes at Lyons, Kansas, it was local geologists who raised questions about the safety of the scheme, while it was an Academy committee, operating with incomplete information, that generally endorsed the AEC's plans.

We do not want to overstate the case. There have been Academy reports that were farsighted, perhaps even ahead of their time. One such as a 1973 report on Western coal lands which warned that there is not enough water in the Western coal states to support the enormous concentration of energy conversion plants projected by utilities and oil companies. That report made Congress aware of the importance of water as a limiting factor in coal development; it appears to have influenced the drafting of strip-mining legislation.[2] Another was a 1962 report on natural resources which warned that domestic crude oil production in the United States might peak and decline within ten years, a prediction which appears to have come true.[3] That report was probably ahead of its time in the sense that its implications went unheeded until the recent Arab oil embargo made the nation aware that it was highly dependent on foreign energy sources.

But the number of farsighted reports is limited. In the 1950s the Academy did little to alert the nation to the dangers of nuclear fallout; that job was left to such crusaders as Nobel chemist Linus Pauling and the small band of scientists who founded the St. Louis Committee for Nuclear Information. In the late 1960s, the Academy made no contribution to the great national debate over whether to build an antiballistic missile system. The lead role in challenging the ABM was played by a group of scientists headed by Jerome B. Wiesner, now president of the Massachusetts Institute of Technology, and the lead role in defending the ABM was played by the Defense Department; the Academy meanwhile contented itself with offering routine technical advice on the hardening of ABM missile sites and on data processing techniques for controlling the system. Similarly, when individual scientists such as Matthew S. Meselson, John Edsall, and E. W. Pfeiffer were raising doubts about U.S. preparations for chemical and biological warfare, the Academy made no contribution to the ensuing debate; it merely continued to help the Defense Department select bright young scientists to work at CBW installations. Nor has the Academy played a lead role in nuclear reactor safety, automobile safety, the environmental movement, or the improvement of health care (though with the formation of the Institute of Medicine the Academy's performance in medical areas is expected to improve significantly). Even President Handler acknowledges that "It might well be argued that the Academy is derelict in failing to mount broad-gauged studies addressed to major domestic concerns of high technical content."[4]

There are many reasons for the Academy's inertia—fear that the Academy's reputation might be muddied if it gets involved in controversial political issues; the drive for consensus which leads to bland, middle-of-the-road reports; the tendency to recommend more research as the solution to complex problems, a prescription that leaves the status quo untouched while study goes forward; reluctance on the part of some scientific leaders to oppose the government too vigorously lest they lose their potential ability to influence decisions; and a feeling by some Academicians that the Academy could not become more aggressive without abandoning its scholarly traditions.

But the chief reason is undoubtedly the Academy's orientation toward serving rich and powerful institutions that do not generally encourage activism in behalf of reform. The Academy's financial dependence on the government and other outside sources makes it difficult for the Academy to launch studies that might conflict with government policy or otherwise shake up established institutions, for few agencies or institutions are so open-minded that they will finance studies which might undercut their policies (a notable case in point being the Interior Department's refusal to allow the Academy to review the environmental impact of the Alaskan pipeline project). But there are no such impediments when it comes to studies that support the aims of established institutions. Indeed, the Academy is generally required by its charter to assist the government upon request, and it usually finds funding plentiful for projects that the government wants performed. It has a long history of advising the Defense Department on such matters as undersea warfare, mine countermeasures, and ship design, but it has done no significant work on arms control. The chief reason is that the Defense Department has an ample budget to support Academy contracts while the Arms Control and Disarmament Agency is impoverished. Similarly, the Academy has a Highway Research Board that has worked to improve the engineering and design of highways, but the Academy has done relatively little, until recent years, on such problems as mass transit or a balanced transportation network. Again, the chief reason is that funds were readily available through the Highway Trust Fund to support the Academy's highway activities, but there was relatively little money to support work on mass transit.

This dependence on outside financial support—coupled with the chartered obligation to assist the government upon request—has largely reduced the Academy to a job shop carrying out tasks for the government. The Academy's prodigious growth since World War II has been dictated not by any internal logic, nor by any sense of how the Academy can best serve the nation, but rather by such accidents as which government agencies happened to ask the Academy for advice. The ever-proliferating requests from the government have forced the Academy to create a sizable bureaucracy, which must then be sustained by still more government contracts, thereby further dissipating the Academy's energies and rendering it even more dependent on government support.

The unfortunate consequence of this dependency, as science critic Daniel S. Greenberg observed in a 1967 analysis, is that the Academy "tends to be most sensitively attuned to serving the *status quo* in society . . . [it] serves the 'ins' and has relatively little traffic with the 'out's' regardless of what issue is at stake."[5]

The Academy has introduced a number of reforms which should help to alleviate some of the defects we have found. These reforms, most of which were discussed earlier in the text, include:

1. the review system, which has unquestionably enhanced the competence and objectivity of Academy reports by requiring that they undergo criticism from eminent scientists who have had no hand in preparing the report;

2. the confidential "bias statement," which seeks to identify factors that might distort the judgment of scientists who have been proposed for membership on Academy committees;

3. the designation of a staff member in the Academy President's office to monitor the appropriateness of committees' appointments and to encourage all units of the Academy to make greater use of young scientists, women, minorities, and other scientists who have not traditionally been part of the Academy's "buddy system";

4. the policy of rotating members off committees after a specified term of service, which is helping to inject fresh viewpoints into committees long dominated by cabals;

5. the visiting committees that are being sent into hitherto autonomous units of the Academy to assess their performance, again injecting fresh viewpoints;

6. the new contract policy under which the Academy retains the right to publish its unclassified reports even if the government wants them suppressed (all such reports, whether published or not, are now supposed to be available for public inspection);

7. the $100,000 fund which was created in 1970 to support projects deemed important by the Academy itself rather than by its clients, thereby providing a small but long overdue base from which to launch independent studies; and

8. the extensive reorganization of the NRC launched under Handler's leadership.

The reorganization plan, which is Handler's major innovation, seeks to improve the quality of the Academy's advisory work by providing closer management supervision. The NRC has traditionally been structured along disciplinary lines. There are divisions devoted to behavioral sciences, engineering, medical sciences, physical sciences, and so forth. But this traditional organization, in Handler's opinion, is not well equipped to cope with major societal problems, such as pollution, poverty, and urban decay, which require the help of experts from a variety of disciplines. Consequently, Handler is remodeling the NRC into two types of functional unit. One type, known as assemblies, will continue to be organized along disciplinary lines and will be concerned with such questions as whether physics is getting adequate financial support and enough trained manpower. The other type, known as commissions and boards, will handle most of the multidisciplinary advisory work; they will deal with public issues in such broad areas as natural resources, human resources, societal technologies, peace and national security, and international affairs. However, the distinction appears blurred in practice; both types of unit are performing somewhat similar advisory chores.

These new units will be run by executive committees of scientists, serving on a part-time basis, who will assume much of the burden for appointing committees and assuring quality control that now "almost overwhelms" the President and NAS Council.[6]

The reorganization could have two beneficial effects on the Academy's performance. The new commissions and boards, which are to encompass a wide range of expertise, may be more broadly competent than the units they replace. And the decentralization of responsibility for quality control may insure more detailed attention to committee appointments and the handling of projects.

But at this writing, it is too early to assess whether the reorganization will actually produce these beneficial results or will merely constitute a meaningless shuffling of boxes on the Academy's organization chart. The chief working unit of the Academy will remain the committee of experts serving on a part-time basis. The reorganization primarily affects the framework upon which the committees are hung.

These reforms are all admirable in intent, and Handler

deserves credit for implementing them despite considerable grumbling from those interested in preserving the old ways of operation. But many of the individual reforms do not go far enough toward eliminating the specific problems they were designed to prevent. And, considered as a whole, *the reforms do not provide a fundamental solution to the Academy's major weakness: its servant–master relationship to the government agencies and industrial interests which provide financial support.*

The surest way to eliminate that problem would be to end the Academy's financial dependence on agencies and industries with special interests in the outcome of Academy investigations. But the Academy has not chosen this admittedly difficult approach. Instead, it has instituted measures designed to insure that, given the fact of the Academy's financial subservience, all reports are done as objectively as possible under the circumstances. The reforms are analogous to the antipollution devices being developed by the automobile companies to reduce pollutant emissions from automobile exhausts. In the case of the automobiles, most engineers would probably agree that the internal combustion engine is inherently dirty and that the surest way to eliminate pollution problems would be to design a new power plant that would be inherently clean. But the automobile industry has opted to stay with the internal combustion engine and is trying to clean up the emissions from these engines by designing antipollutant devices that will remove the noxious gases before they are released into the atmosphere. The situation is inherently unstable. If the devices fail, the engine reverts to its dirty state and emits pollutants again. Much the same can be said of the Academy's recent measures. They leave untouched the master–client relationship, and simply try to clean up the Academy's products.

We offer no panaceas, no perfect cure for the ills we have diagnosed. But we believe the Academy could take several steps to enhance its independence, objectivity, and social utility. Some steps would require major change; others would be modest extensions of the reforms already implemented. In making these proposals, we are forced to walk a difficult tightrope. We believe that the Academy's *potential* independence of government, industry, and other powerful interests is one of its most precious assets. We are thus reluctant to see compulsion applied to the Academy, because compulsion necessarily erodes that independence. On the other hand, we believe that the Academy's work is potentially (and in some cases actually) of such great importance to the nation that further reform is required. The best solution would be for that reform to spring independently from within. But if the Academy fails to correct its weaknesses, then some form of compulsion from without, perhaps by Congress, may be necessary. Although Congress has no direct authority over the Academy's internal operations, it could pass legislation stipulating that no government agency could contract for an Academy study unless the Academy institutes meaningful reform.

The Academy could insure greater competence, objectivity, and credibility for its advisory reports by soliciting participation of the activists it traditionally shuns and by opening its advisory processes to outside scrutiny. Specifically, we recommend:

Acceptance of a Task

The Academy should announce each project it undertakes so that a broad range of the scientific community and the public knows that an Academy study is about to get under way. As it is now, the Academy and its funding agencies announce only a handful of projects at the time they are initiated; thus few people are aware of any but the most visible Academy projects until a final report is issued at the completion of the project. Early announcement would allow interested scientists to nominate possible committee members and submit relevant data and would make it difficult for the government or the Academy leadership to bury a report should they feel inclined to violate recent assurances that such suppression will no longer occur. The announcements could be made through press releases, the Academy's monthly newsletter, its scientific journal, and any other medium which reaches a significant audience. Government funding agencies should similarly announce every contract they sign with the Academy.

Selection of Committee

The Academy should cast its net wider in selecting scientists to serve on its advisory committees, for the choice of committee members is probably the most crucial factor influencing the quality of a report. The current "buddy system," which allows a handful of staffers and scientists to generate the names for a committee, results all too often in committees which lack relevant expertise or are biased toward certain viewpoints.

One possible improvement, which has been called "worthy of trial" by one of the Academy's own reports, would be to allow scientists to nominate themselves or their colleagues, perhaps in response to advertisements in professional journals announcing that a committee was about to be formed.[7] That approach would presumably draw volunteers who are strongly motivated and knowledgeable about the subject in question, but who might be unknown to the Academy or else out of favor with the small circle generating names.

Another possible approach would be to appoint an ombudsman to scour the land in search of scientists with adversary views on a subject, much as a journalist digs hard to find opposing viewpoints on an issue. It is our observation that Academy committees tend to shun the activists on any issue on the theory that they are biased or troublesome; but they show fewer qualms about appointing scientists from the "Establishment" side, apparently perceiving less bias from that quarter. Yet, when pressure is applied, the Academy can generally find scientists of independent or even activist inclinations who satisfy its criteria for appointment. This became apparent in 1970 when an Academy committee was formed to assess the biological effects of ionizing radiation. The committee initially included none of the scientists who had challenged existing radiation standards, but it did include scientists who had defended the standards against the critics. After complaints by Senators Edmund S. Muskie (D-Maine) and Mike Gravel (D-Alaska), the Academy added several scientists to the committee who were recognized as neutral or even critical toward radiation policy.[8] The resulting report did not fully support the critics, but it concluded that the existing standards were "unnecessarily high" and could result in about six thousand additional cancer deaths per year.[9] Although it is impossible to be certain what the original committee would have concluded, it seems likely that the report turned out more critical simply because the Academy was induced, in this instance, to try harder than usual to find responsible critics to appoint to the committee.

Elimination of Bias

Academy committees which include industry scientists should counterbalance them with qualified scientists from citizen, consumer, and public interest groups. Similarly, Academy units that have liaison panels with industry should establish parallel panels representing the public interest. Industry expertise is unquestionably useful to the deliberations of many committees, but it should not be allowed to exert unchallenged sway over the thinking of a committee.

The Academy should also make public the bias statements submitted by its committee members. These statements reveal a scientist's job history, consulting relationships, financial interests, sources of research support, previous public stands on an issue, and other information bearing on his objectivity and independence. Lacking such information, it is virtually impossible for an outsider to begin to judge whether a particular Academy committee is well-balanced or biased. Academy officials sometimes contend that scientists would refuse to serve on Academy committees if their bias statements were public. But the number of refusals would probably not be large, and an organization that professes the highest objectivity in dealing with issues of great public importance should be willing to demonstrate that objectivity by full disclosure. The bias statements should also be required of all staff members, for they, too, are in a position to influence the conclusions of Academy reports.

Gathering of Information

Academy committees should establish a mechanism for actively soliciting adversary views. Under the present system the committees are deluged with data from well-organized, well-financed agencies and corporations, but they are less apt to receive the views of feebly supported dissenting groups.

The Committee on Motor Vehicle Emissions, for example, received extensive information from the automobile companies. But, when it solicited comments from the public by sending letters to some seven hundred organizations and individuals and by placing a statement in the *Federal Register*, it found the response was "disappointingly small."[10] Such passive quests for divergent views are not enough; the Academy should assume active responsibility for ferreting out all relevant dissenting viewpoints.

One possible approach, suggested in a different context by the Academy's panel on technology assessment, would be to create "surrogate representatives" for the public.[11] These representatives would be responsible for digging out relevant information regardless of origin, and for speaking on behalf of nonorganized interests. A parallel approach, also suggested in another context by Academician Arthur R. Kantrowitz, vice-president and director of the Avco Corporation, would be to conduct adversary proceedings before Academy committees. Scientist advocates could be appointed to represent differing points of view and could cross-examine each other in front of the experts on the committee, thus insuring a fair hearing for all sides of a controversial issue.[12]

Conduct of Committee Business

The Academy should open its advisory proceedings to public inspection. As it stands now, Academy committees meet behind closed doors and they keep most of their records hidden. Only the contracts which initiate a project and the final reports which result from the project are generally available for outside scrutiny. Everything else—the criteria used for selecting the committee, the bias statements filed by committee members, the minutes and correspondence of the committee, the data on which it based its conclusions, and the internal reviews of committee reports—is considered privileged information. Thus it is extraordinarily difficult to monitor the performance of an Academy committee. Even experts find it hard to evaluate a committee's judgment without access to the original data on which that judgment is based. And the public has no way to judge whether special interests have influenced the committee's deliberations. The excuse generally given for this secrecy is that the committee members can speak more frankly, and with less fear of outside pressure, if the public is barred. But if the process were opened up more—if all meetings and records were made public—then the committees might benefit from unexpected insights volunteered by the interested public, and the public could assure itself that the project was being well handled.

The Academy contends that it is not governed by the Federal Advisory Committee Act—legislation that was adopted in 1972 in an effort to open the government's advisory apparatus to public inspection.[13] The House–Senate conference report explaining that act states that it "does not apply to persons or organizations which have contractual relationships with federal agencies nor to advisory committees not directly established by or for such agencies."[14] And supporters of the act explained in floor debate that this provision specifically excluded the Academy.[15] We agree that it is beneficial that the Academy should not be subject to the act as a whole because it contains provisions that would seriously erode the independence of the Academy. The act provides, for example, that no advisory committee can meet without advance approval from, and attendance of, a designated government official; that the agenda must be approved by that official; and that he may adjourn a meeting whenever he deems it "in the public interest."[16] Such provisions would give the government funding agencies much greater power than they now possess to influence a committee's deliberations. But other provisions of the act should apply to the Academy. These include requirements (with certain exceptions) that meetings be open to the public, that interested persons be permitted to file statements and appear before the committee, that detailed minutes be kept and made public, and that "The records, reports, transcripts, minutes, appendices, working papers, drafts, studies, agendas, or other documents which were made available to or prepared for or by each advisory committee shall be available for public inspection and copying."[17] *The rationale for applying these provisions to the Academy is that the Academy, in its work for the government, exerts influence over public policy and serves essentially the same function as the government's own in-house advisory committees.* We believe the Academy should voluntarily accept most of the provisions of the act. If the Academy lags in complying, then Congress should consider amending the act to make the relevant provisions apply to the Academy's contractual work for the government. Similarly, the 1966 Freedom of Information Act, which is incorporated by reference in the advisory committee act, should apply to the Academy just as it does to government agencies.*

Although the Academy claims to be a "private" organization, it would apparently not dispute the power of Congress to require more openness in Academy procedures. When we asked the Academy's law firm, "In your opinion, does the Congress have the constitutional power to subject by amendment the National Academy of Sciences and its constituent groups to the Freedom of Information Act or the Federal Advisory Committee Act?"[18] the lawyers replied, "Presumably it could, but only by an explicit act such as amendment of the Academy's charter."[19]

Preparation and Publishing of Reports

The Academy should increase the impact and usefulness of its reports by requiring its committees to explain the potential significance of their findings for public policy.

Harvard law professor Milton Katz has suggested that Academy advisory groups should develop "an accepted practice" under which they "regularly take pains to sort out, recognize, and identify ... whether they are rendering an objective assessment or advocating a cause, and whether they are speaking as experts within their field of special knowledge and competence or as citizens concerning a question of general public policy."[20] Applying this sound advice

---

*A suit filed against the Academy in 1974 by a public interest group advanced the novel argument that the Academy should be deemed a government "agency" for purposes of the Federal Advisory Committee Act, while its committees should be considered advisory groups answerable to the Academy itself rather than to the funding agencies. Under that theory, the Academy would maintain administrative discipline over its own committees. Such a finding would open Academy committees to public inspection while avoiding most of the dangers of increased control by the funding agencies. But it might also subject the Academy to influence from the Office of Management and Budget, which supervises agency use of advisory committees.

to our case history material, we would argue that the Academy's SST–sonic boom committee should have expressed its professional belief that the sonic boom would be unacceptably annoying to people; and that, acting as a group of informed citizens, it should have rendered a judgment as to what that finding implied for the future of the SST project. Similarly, the committee which examined airborne lead should have explicitly stated its "professional" opinion of the degree of hazard and its "citizen" opinion as to whether lead should be removed from gasoline.

The Academy should further enhance the value of its reports by publishing viewpoints that diverge from the consensus. To begin with, the Academy should encourage dissenting opinions by committee members who feel uneasy with the majority view. Some committees already do this, but it is our conclusion that most try too hard for unanimity, with the result that important differences are papered over and bland reports are produced. If the Academy wants to pursue its Supreme Court aspirations to the end, it should remember that some of the most brilliant and foresightful opinions in Supreme Court history have been dissenting opinions.

Second, the Academy should publish the opinions of the eminent scientists who have been asked to conduct internal reviews of committee reports. If the reviewers agreed with the authoring committees, their opinion would give added weight to the report. And, if they disagreed, their comments would alert public officials to possible defects in the report. We do not mean to imply that the reviewers are always right in their judgments, but they are generally as well qualified as anyone to assess the merits of a report. Public debate and decision-making would benefit if every Academy report included an appendix summarizing the reviewers' opinions of the final report and delineating the points of disagreement.

The report should also include detailed information on the backgrounds of all committee members, advisers, and staffers who participated in its preparation, including the qualifications which led to their selection for the project and the activities revealed on their bias statements. Currently the Academy simply lists the names and primary job titles of all committee members at the front or back of a report. Thus a professor of chemistry serving on a pollution committee is simply listed by his academic title; there is no indication as to whether he has ever done any work relating to the pollutant under examination or whether he has consulted for industrial polluters. From the information on committee members now published in Academy reports, it is virtually impossible for an outsider to judge whether the committee was well balanced and possessed the necessary expertise.

Our last suggestion concerning reports is that all except those bearing a security classification should be public. Currently, many Academy committees submit memoranda and advisory letters to government agencies that are not published as Academy reports. The existence of these documents is seldom made known, and some committees refuse to release them upon request. We cited the case of the NAE Aeronautics and Space Engineering Board, which submits privileged reports to the National Aeronautics and Space Administration concerning such controversial projects as the space shuttle and the supersonic transport. Similarly, a joint NAS–NAE Advisory Committee to the Department of Housing and Urban Development (HUD) submitted two dozen technical reports in 1971–1972 that were not available for public inspection until an investigator for the Center for Study of Responsive Law successfully demanded that HUD release them in accord with the Freedom of Information Act. The Academy's information office asserts that, under a policy in effect since 1972, all unclassified reports are supposed to be open for public inspection. But the word appears not to have filtered down throughout the Academy bureaucracy.

The best way to create an independent Academy—one free to use its unique capabilities to attack problems of urgent importance and prepared, if necessary, to adopt an aggressive adversary stance toward powerful agencies and institutions—would be to reorient the Academy in its relationships with other organizations.

For most of its recent history, the Academy has thought of itself as a consultant serving the needs of its client agencies. That relationship works well when the interests of a government agency and the interests of the public coincide, but when the interests diverge the Academy often subordinates its doubts about government policy so as to remain an effective part of the government "team." The bureaucracy and the committees of the Academy still tend to think of themselves as cohorts of the contracting agencies. Some staff members told us that there is no difference between the public interest and the government's interest; others refused to show us correspondence and reports to government agencies unless the agencies agreed to release them; and many told us that the Academy has no business looking into agency programs unless the agency requests help. As one staffer expressed it: "No one has appointed the Academy a national ombudsman."

We believe that attitude is short-sighted. The Academy should designate itself a national ombudsman—much as the press in this country has appointed itself guardian of the public interest—and every Academy committee should approach its task as if it were representing the public rather than offering consulting services to a particular agency. The best Academy committees already do this, but far too many take the narrower approach. The Academy has often been willing to take tough stands in defense of the scientific community—as when it protested Soviet harassment of physicist Andrei Sakharov even though the Nixon Administration wanted to downplay the incident.[21] Now the Academy should become equally outspoken on behalf of the general public.

The new outlook should be accompanied by a broadening of the influences which determine the Academy's agenda. Traditionally, the majority of Academy projects are undertaken at the request of a federal agency, while a minority are conceived by the Academy itself which then solicits support from the government or another outside source. The result of the agencies' dominance, as we have seen, is that the Academy is brought into action in support of agency goals, but it is not asked to conduct studies that might conflict with those goals.

The power of the agencies to command Academy assistance should be reduced. The charter seemingly requires the Academy to answer virtually every government request, no matter how trivial or what motives lie behind it. However, the Academy occasionally turns aside a request and it often changes the thrust of the questions asked. The

Academy should become much bolder in declining to accept tasks where the ground rules are defined narrowly by the contracting agency. If necessary, the charter should be amended to make it clear that the Academy is free to decline requests.

At the same time, the Academy should open itself to requests from a broader constituency. To some extent, this is already happening. Congress, the most broadly constituted arm of government, has increasingly sought Academy advice in recent years, and the newly created Congressional Office of Technology Assessment is expected to rely frequently on the Academy for advisory studies. But the Academy should seek an even broader infusion of ideas by setting up a mechanism whereby any citizen or scientist could suggest suitable projects for the Academy. In particular, the Academy should solicit ideas from activist organizations— such as the Federation of American Scientists, the Scientists Institute for Public Information, the Union of Concerned Scientists, the Center for Science in the Public Interest, and the Environmental Defense Fund—which have often been far ahead of the Academy in sensing technology-related problems and pushing for reform.

The Academy will never become truly independent until it reduces its reliance on outside financial sources. There has been considerable discussion within the Academy on how to obtain more "free money" to support projects that the government is unwilling to fund. But none of the alternatives seems promising. Although Handler warned the members that "The weakness of our financial underpinning is a serious deterrent to assuring the continuing quality and credibility of all aspects of the operation," he was not sanguine about finding a remedy: "I see little hope of securing endowment on a scale sufficient to repair this intrinsic defect in our structure. The large philanthropic foundations exhibit decreasing tendencies to provide bloc funding for institutional support, although they do regard us as a favored customer when seeking specific project support."[22]

Handler said he would explore other funding alternatives, but he seems not to have given serious consideration to a radically different solution—reducing the scale of Academy operations so that endowment, both that which is in hand and that which might be raised, could support a substantial part of the remaining structure. The goal should be to reduce the permanent staff to a size that can be supported by the Academy's own funds, thus obviating the scramble for government projects to keep the staff alive. If necessary, the Academy could supplement this core staff by drawing talent from the scientific community on a short-term basis to staff a particular project with the understanding that such appointees would return to their other pursuits upon completion of the project. These short-termers would have no need to grovel for follow-up contracts.

Such a scaling down of operations is not a new idea; it has been talked about for years. Handler himself has acknowledged that "A large number, if not the majority, of members are aghast at the range of activities of the National Research Council" and have expressed "a strong feeling that the Research Council should be markedly reduced in size and scope." Nevertheless, Handler added, "Meeting after meeting, the Council and I agree to yet more projects which lead to yet more committees. While the propriety of some of these new projects concerns me, yet more serious may be the older projects and committees which to a considerable extent have taken on a life of their own."[23] But all this talk has led to very little action, for resistance develops whenever Academy leaders try to eliminate any particular activity. Narrow interests within the Academy and in the funding agencies inevitably claim that the activity is valuable and that the nation will suffer if the Academy gives it up. The problem clearly needs to be attacked on a broad front rather than piecemeal, and courageously rather than timidly. A high-level decision should be made to pare off all activities that do not require the unique capabilities of the Academy. The criteria for undertaking a project should be that it is of great importance, that it demands the highest technical competence, and that it requires independence of judgment.

By this yardstick, the Academy *would not* advise the government on research needed to understand the sonic boom, because such a task did not require independent judgment and could be performed by any competent technical committee. However, the Academy *would* advise the government on the sonic boom's likely impact on the citizenry and on the SST's likely effect on the atmosphere, because such ques-

tions require a more independent look than they are apt to
get from the Department of Transportation, which is promoting the SST. Similarly, the Academy *would not* perform
a routine safety evaluation of the food coloring Red 2
because the Food and Drug Administration could easily
perform that evaluation on its own. But the Academy *would*
advise the government on radioactive waste disposal policies because that problem is highly complex, it involves
potentially staggering consequences for future generations,
and it requires a more independent appraisal than the
Atomic Energy Commission, which promotes nuclear power,
is apt to give it. (Unfortunately, the Academy's performance
on the radioactive waste issue was far from independent, as
we have noted.)

There would undoubtedly be disagreement over whether
particular projects met the criteria, but the important point
is to shift the burden of proof. Instead of performing every
service requested unless it can think up a reason for declining, the Academy should decline every project unless it can
justify its undertaking. The Academy likes to style itself a
"Supreme Court of Science," but it seems to have forgotten
that the actual Supreme Court turns down more cases than
it hears. No one believes that every traffic case should be
resolved by the highest court in the land.

The Academy could never become, and probably should
never become, completely independent of outside funding.
But, if the Academy's endowment were more nearly equal to
supporting the resident staff, then the Academy could be
much more selective about the funding arrangements it is
willing to accept. It could decline all industry funding for
committees which consider issues crucial to industry. And it
could refuse to accept funding from government agencies
that have an institutional bias in the issue to be investigated. The appropriate source of funding might vary from
issue to issue, but some of the "cleanest" sources would
include the new Office of Technology Assessment, the
National Science Foundation, philanthropic foundations,
and Congress as a whole. Although individual Congressional
committees often have strong biases on an issue, Congress
as a whole includes so many conflicting viewpoints that it is
not apt to pressure the Academy to reach a particular verdict.

Our proposals have ranged from those which would be easy to effect (opening the committee meetings) to those which would cause major disruptions in the Academy's traditional structure (radical surgery on the scale of operations). None of our recommendations guarantees that the Academy will show greater initiative, or perform at a higher level of quality, or better protect the public interest. But, taken as a whole, we believe our recommendations would help boost the Academy's performance toward the highest possible standard. If that goal is achieved, then the Academy might come to deserve the accolade—bestowed by one of its former presidents—that it is "a supreme court of final advice" whose findings, insofar as is humanly possible, are "wholly in the public interest uninfluenced by any elements of personal, economic or political force."[24]

NOTES

PREFACE

1. W. H. Bradley to Philip M. Boffey, August 7, 1971.

2. The three most comprehensive journalistic analyses of the Academy are D. S. Greenberg's three-part series in *Science*, vol. 156, pp. 222, 360, 488 (1967); Claude E. Barfield's two-part series in *National Journal*, vol. 3, pp. 101, 220 (1971); and John Walsh's two-part series in *Science*, vol. 172, pp. 242, 353 (1971).

3. The policy is referred to in a letter from Ernst Weber, chairman of the Academy's division of engineering, to Louis V. Lombardo, president of the Public Interest Campaign, February 1972.

4. The petition is quoted in the *Newsletter* of the National Association of Science Writers, vol. 19, no. 2, June 1971.

5. *Clean Air*, the Public Interest Campaign's monthly Washington report, Vol. III, Nos. 3 & 5, March–April–May 1974, p. 1.

6. Memorandum from Philip Handler to Heads of Major Academy Offices, September 17, 1973.

7. National Academy of Sciences, *The Science Committee*, Washington, D.C., 1972, p. iii.

8. Folke Skoog to Philip M. Boffey, August 11, 1971.

9. Samuel Silver to Philip M. Boffey, September 27, 1971.

10. J. R. Pierce to Philip Handler and Clarence Linder, August 12, 1971.

11. Albert Szent-Györgyi to Philip M. Boffey, August 17, 1971.

12. Olin C. Wilson to Philip M. Boffey, August 26, 1971.

13. Walter M. Elsasser to Philip M. Boffey, November 10, 1971.

14. Field Winslow to Philip M. Boffey, August 23, 1971.

CHAPTER ONE

1. Quoted in Jerome B. Wiesner, "John F. Kennedy: A Remembrance," *Science*, vol. 142, no. 3596, November 29, 1963, p. 1148.

2. Excerpts from Udall's speech were published in the *New York Times*, December 31, 1970, and *National Journal*, vol. 3, no. 3, January 16, 1971, p. 111.

3. "Academy Critic Shoots Wildly But Hits Sore Spot," *Nature*, vol. 229, January 15, 1971, p. 151.

4. Handler's letter, which was not published by the *Times*, was printed in the NAS *Letter to Members*, vol. 1, no. 5, January 1971. This is a "private communication" not available to outsiders.

5. See note 3 above.

6. John Walsh, "National Research Council (II): Answering the Right Questions?" *Science*, vol. 172, April 23, 1971, p. 356.

7. *Report of the National Academy of Sciences for the Year 1863*, Washington, D.C., 1864, p. 2.

8. Quoted in D. S. Greenberg, "The National Academy of Sciences: Profile of an Institution (I)," *Science*, vol. 156, April 14, 1967, p. 222. Copyright © 1967 by the American Association for the Advancement of Science.

9. A. Hunter Dupree, *Science in the Federal Government*, The Belknap Press of Harvard University Press, Cambridge, Massachusetts, 1957, pp. 135–140.

10. Joseph Henry to Stephen Alexander, March 9, 1863, quoted in Nathan Reingold, ed., *Science in Nineteenth Century America: A Documentary History*, Hill and Wang, New York, 1964, p. 204.

11. An Act to Incorporate the National Academy of Sciences, approved March 3, 1963, reprinted annually in the Academy publication entitled *Organization and Members*.

12. National Research Council, "Information for Members of Divisions, Committees, Boards, and Panels," pamphlet dated July 1, 1969.

13. President's Report, NAS Business Meeting, April 23, 1968.

14. Frank B. Jewett, "The Academy—Its Charter, Its Functions and Relations to Government," read before the Academy, November 17, 1947, published in *Proceedings of the National Academy of Sciences*, vol. 48, no. 4, April 15, 1962.

15. Dupree, *Science in the Federal Government*, pp. 146–147.

16. I. Bernard Cohen, "Some Reflections on the State of Science in America During the Nineteenth Century," *Proceedings of the National Academy of Sciences*, vol. 45, 1949, p. 671.

17. F. W. True, *A History of the First Half-Century of the National Academy of Sciences, 1863–1913*, National Academy of Sciences, Washington, D.C., 1913, Appendix IX.

18. R. A. Millikan, *Autobiography*, 1950, p. 132.

19. Helen Wright, *Explorer of the Universe: A Biography of George Ellery Hale*, New York, Dutton, 1966, pp. 185–186.

20. The founding of the NRC is described in detail in Dupree, *Science in the Federal Government*, pp. 308–330.

21. D. S. Greenberg, "The National Academy of Sciences: Profile of an Institution (II), *Science*, vol. 156, p. 361.

22. Walter Goldschmidt, "Equinoxial Rites of the National Research Council," *Science*, vol. 174, no. 4008, October 29, 1971, pp. 474–476.

23. Alfred S. Romer to Philip M. Boffey, August 18, 1971.

24. National Academy of Sciences, *Science and Technology in Presidential Policymaking: A Proposal*, Washington, D.C., June 1974, p. 31.

25. National Academy of Sciences, *The Science Committee*, Washington, D.C., 1972, App. D, Table D-1.

26. *Letter to Members* (a private communication), June 1971.

27. Quoted in Ronald W. Clark, *Einstein: The Life and Times*, World Publishing Company, New York and Cleveland, 1971, p. 389.

28. Steve Smale, "The Annual Meeting of the National Academy of Sciences," *New York Review*, July 1, 1971, p. 38.

29. Britton Chance to Philip M. Boffey, August 12, 1971.

30. Report of the Managing Editor to the Editorial Board of the *Proceedings of the National Academy of Sciences*, March 30, 1971.

31. See Barbara J. Culliton, "NAS: A Face-lifting for the Proceedings," and "Academy Turns Down a Pauling Paper," *Science*, vol. 177, no. 4047, pp. 408–409ff. Letters commenting on the articles were published in *Science*, vol. 177, no. 4055, p. 1152, and vol. 177, no. 4062, p. 696.

32. Philip Handler, "Scientific Relationships Between the United States and the Soviet Union," testimony before the subcommittee on International Cooperation in Space and Science of the Committee on Science and Astronautics, House of Representatives, July 14, 1972.

CHAPTER TWO

1. S. Cole and J. R. Cole, *Amer. Sociol. Rev.*, vol. 32, no. 377 (1967), quoted in Don E. Kash, et al., "University Affiliation and Recognition: National Academy of Sciences," *Science*, vol. 175, no. 4026, p. 1077.

2. Organization for Economic Cooperation and Development, *Reviews of National Science Policy: United States*, Paris, 1968, Paragraph 560.

3. *Science & Government Report*, vol. 1, no. 19, p. 2.

4. NAS *Proceedings*, April 1878, pp. 132–133.

5. Richard Carter, *Breakthrough: The Saga of Jonas Salk* (New York, Trident Press, Inc., 1966), p. 299.

6. *Ibid.*, p. 299.

7. *Science & Government Report*, vol. 1, no. 7, May 1, 1971, p. 1.

8. R. W. Sperry to Philip M. Boffey, August 27, 1971.

9. Don E. Kash, Irvin L. White, John W. Reuss, and Joseph Leo, "University Affiliation and Recognition: National Academy of Sciences," *Science*, vol. 175, no. 4026, March 10, 1972, pp. 1076–1084. Copyright © 1972 by the American Association for the Advancement of Science.

10. Irvin L. White, Don E. Kash, and John W. Reuss, "University Affiliation and Recognition: A Comparison of Two Elites in American Science," unpublished manuscript.

11. Kash, et al., *Science*, March 10, 1972, p. 1083.

12. The members of the various disciplinary sections of the NAS are listed in the publication, *National Academy of Sciences of the United States of America, Membership*, July 1, 1971.

13. National Science Foundation, *American Science Manpower 1970, A Report of the National Register of Scientific and Technical Personnel*, NSF 71-45, U.S. Government Printing Office, Washington D.C., 1971; Table A-13, p. 82.

14. *Ibid.*, Table A-6, pp. 49, 50.

15. Stanley A. Cain to Philip M. Boffey, August 10, 1971.

16. D. M. Hegsted to Philip M. Boffey, August 18, 1971.

17. S. A. Goudsmit to Philip M. Boffey, August 23, 1971.

18. The age breakdown is in *Letter to Members* (a private communication), vol. 1, no. 7, March 1971, p. 12.

19. *Letter to Members*, vol. 1, no. 8, June 1971, p. 13.

20. E. N. Parker to Philip M. Boffey, September 9, 1971.

21. National Academy of Sciences, "Minutes of the Annual Meeting," April 27 and 28, 1971, p. 6.

22. *Letter to Members*, vol. 2, no. 2, September 1971, p. 7.

23. *Letter to Members*, vol. 2, no. 8, May 1972, p. 19.

24. Annette Cronin to Philip M. Boffey, January 4, 1972.

25. Dwight J. Ingle to Philip M. Boffey, August 16, 1971.

26. "Minutes of the Annual Meeting," April 27 and 28, 1971, p. 4.

27. See, for example, Lamont C. Cole, "Can the World Be Saved?" (an article adapted from an address before the American Association for the Advancement of Science), *New York Times Magazine*, March 31, 1968, pp. 34–35ff. The NAS Council's concern is described in "Academy Squirms a Little in Sudden Spotlight," *Nature*, vol. 231, May 7, 1971, p. 6.

28. See, for example, his speech "Science and the Federal Government: Which Way to Go," before the Federation of American Societies for Experimental Biology, Atlantic City, N.J., April 14, 1970, p. 4.

29. Luther J. Carter, "National Academy of Sciences: Unrest Among the Ecologists," *Science*, vol. 159, January 19, 1968, p. 287.

30. Philip Handler, "Toward a National Science Policy," Statement before the Subcommittee on Science, Research, and Development, Committee on Science and Astronautics, House of Representatives, July 21, 1970, p. 2.

31. Philip Handler, "Science and Scientists: Obligations and Opportunities," Sigma Xi Lecture, University of Houston, Houston, Texas, October 21, 1970.

32. Philip Handler, "In Defense of Science," speech before the general meeting of the American Iron and Steel Institute, New York, May 26,

1971. (Virtually identical remarks were made to the International Congress of Pure and Applied Chemistry in Boston, Mass., July 26, 1971.)

33. Handler's retraction was issued through the press office at the Boston convention and was published in the *Washington Post*, July 30, 1971.

34. Arthur W. Galston to Philip M. Boffey, September 9, 1971.

35. W. H. Bradley to Philip M. Boffey, August 7, 1971.

36. *Letter to Members*, vol. 2, no. 7, p. 4.

37. Memorandum entitled "Committees of NAS/NAE/IOM/NRC, Some Statistics," April 1974.

38. Quoted in *Letter to Members*, a private communication, November 1969, p. 11.

39. Philip Handler to Senator Lee Metcalf, December 3, 1971, in U. S. Senate Committee on Government Operations, *Advisory Committees*, Hearings Before the Subcommittee on Intergovernmental Relations, 92nd Cong., 1st Sess., Part 3, October 1971, p. 1033.

CHAPTER THREE

1. National Academy of Sciences, *Constitution*, October 25, 1971, Article II, Sec. 6.

2. Philip Handler to the President of the Senate, the Speaker of the House of Representatives, the Administrator of the Environmental Protection Agency, February 15, 1973, pp. 4, 5. This is the cover letter to National Academy of Sciences, *Report by the Committee on Motor Vehicle Emissions*, Washington, D.C., February 12, 1973.

3. *New York Times*, February 22, 1973.

4. *New York Times*, March 14, 1973.

5. National Academy of Sciences, *The Science Committee*, Washington, D.C., 1972, Appendix E, p. 62.

6. Philip Handler, "Annual Report of the President, National Academy of Sciences," April 23, 1974.

7. *The Science Committee*, Appendix B, p. 27.

8. *Ibid.*, Appendix D. pp. 40, 53.

9. Memorandum entitled "Committees of NAS/NAE/IOM/NRC, Some Statistics," April 1974.

10. *Marihuana and Society*, a joint statement by the Council on Mental Health and the Committee on Alcoholism and Drug Dependence of the American Medical Association and the Committee on Problems of Drug Dependence of the National Research Council. Published in the Academy's *News Report*, vol. 18, no. 6, June–July 1968, p. 3ff. Also published in the *Journal of the American Medical Association*, June 24, 1968.

11. John Kaplan, *Marijuana—The New Prohibition*. The World Publishing Company, New York and Cleveland, 1970, p. 189.

12. National Academy of Sciences, *U.S. International Firms and R, D & E in Developing Countries*, Washington, D.C., 1973, p. xiv.

13. National Research Council. Division of Biology and Agriculture, *Annual Report, FY 1970*, p. 6.

14. National Academy of Sciences, *The Great Alaska Earthquake of 1974, Summary and Recommendations*, Washington, D.C., 1973, p. 108.

15. *The Science Committee*, Appendix B, p. 21.

16. *The Science Committee*, Appendix E, p. 73.

17. See Chapter 9.

18. "Comments on Contracts with the National Academy of Sciences FY 1968–1972," a report from Harry W. Hays, science adviser, to T. W. Edminster, administrator, Agricultural Research Service, United States Department of Agriculture, December 6, 1971.

CHAPTER FOUR

1. Claude E. Barfield, "Handler Moves to Make Academy More Active, Useful as Federal Adviser," *National Journal*, vol. 3, no. 3, January 16, 1971, p. 224.

2. Harvey Brooks, "Can Science Survive in the Modern Age?" *Science*, vol. 174, no. 4004, October 1, 1971, p. 28. Copyright © 1971 by the American Association for the Advancement of Science.

3. Rogers C. B. Morton, Secretary of the Interior, to Philip Handler, March 29, 1971.

4. *Congressional Record*, vol. 118, pt. 17, June 20, 1972, p. 21661.

5. *Letter to Members*, vol. 2, no. 8, May 1972, pp. 6–7.

6. National Academy of Sciences, *Technology: Processes of Assessment and Choice*, a report to the Committee on Science and Astronautics, U.S. House of Representatives, U.S. Government Printing Office, Washington, D.C., July 1969, pp. 59, 61, 62.

7. *Ibid.*, p. 100.

8. *Letter to Members*, vol. 2, no. 6, February 1972, pp. 8–9. For similar sentiments on another issue, see *Letter to Members*, vol. 2, no. 8, May 1972, p. 4.

9. National Academy of Sciences, *Act of Incorporation*, March 3, 1863, Sec. 3.

10. Jewett, "The Academy . . ." p. 485.

11. Frederick Seitz, "President's Report," NAS Business Meetings, April 23, 1968.

12. Philip Handler to R. C. Lewontin, January 22, 1969.

13. R. C. Lewontin to Philip Handler, July 31, 1970.

14. U.S. Senator Gaylord Nelson to the Honorable John H. Chafee, Secretary of the Navy, May 3, 1971.

15. John H. Foster, director of defense research and engineering, to Philip Handler, June 3, 1971. Quoted as the specific task of the committee in National Academy of Sciences–National Research Council, *Summary Statement of the Ad Hoc Panel on Sanguine*, May 1972, p. iii.

16. National Academy, *Summary Statement*, p. 2.

17. Memorandum to G. B. Kistiakowsky from E. M. Purcell, May 9, 1972.

18. Comptroller General of the United States, *Medical Education for National Defense Program*, Report to the Committee on Appropriations, House of Representatives, B-162455, March 14, 1968.

19. U.S. Congress, House Committee on Appropriations, *Department of Defense Appropriation Bill, 1969; Report*, 90th Cong., 2nd Sess., July 18, 1968, p. 31.

20. Quoted in National Academy of Sciences, *Final Report of the Ad Hoc Committee on the Medical Education for National Defense Program*, May 22, 1969, p. 1.

21. National Academy of Sciences, *News Report*, vol. XVII, no. 3, March 1967, p. 7.

22. *Final Report of the Ad Hoc Committee*, Part I, p. 7.

23. Charles L. Dunham to Louis M. Rousselot, February 11, 1969, reproduced in *ibid.*, Appendix D.

24. U.S. Congress. House Committee on Appropriations. *Department of Defense Appropriations for 1970. Part 2. Operation and Maintenance.* Hearings before a subcommittee. 91st. Cong., 1st Sess., p. 411.

25. National Academy of Sciences–National Research Council, *Resources and Man*, W. H. Freeman and Company, San Francisco, 1969, p. 10. The helium conservation program and its problems are discussed in Philip M. Boffey, "Helium: Costs Jeopardize Future of Government Conservation Program," *Science*, vol. 167, March 20, 1970, pp. 1593–1596; and Charlotte A. Price, "On the Future of Helium," paper reprinted in *Congressional Record*, vol. 118, pt. 14, 92nd Cong., 2nd Sess., May 16, 1972, pp. 17497–8.

26. *Articles of Organization of the National Academy of Engineering*, October 1970, Article V.

27. "Merit and Power," *Nature*, vol. 230, April 30, 1971, p. 549.

28. *Letter to Members*, vol. 1, no. 8, June 1971, p. 3.

29. Allen L. Hammond, "Academy Says Energy Self-Sufficiency Unlikely," *Science*, vol. 184, no. 4140, May 31, 1974, p. 964. Copyright © 1974 by the American Association for the Advancement of Science.

30. National Academy of Engineering–National Research Council, *Abatement of Sulfur Oxide Emissions from Stationary Combustion Sources*, Washington, D.C., 1970, p. 3.

31. Statement of Charles F. Barber before Puget Sound Aid Pollution Control Board, July 8, 1970. Taped excerpts of meeting supplied to

study team by William H. Rodgers, Jr., professor of law, University of Washington.

32. *New York Times*, October 22, 1970.

33. *Salt Lake Tribune*, June 24, 1970.

34. U.S. Department of Health, Education, and Welfare, *Transcript of Proceedings of Executive Session of Abatement Conference of Parkersburg, West Virginia*, Vienna, West Virginia, November 13, 1970, p. 28.

35. NAE, *Abatement of Sulfur Oxide Emissions*, p. 34.

36. See Metcalf's statements appended to U.S. Congress, Senate Committee on Interior and Insular Affairs, *Magnetohydrodynamics (MHD)*, Hearing before the Subcommittee on Minerals, Materials, and Fuels, 91st Cong., 2nd Sess., February 23, 1970, Part II, pp. 193ff. The NAE committee's statements about rate structure are specifically criticized in two articles by Richard Gilluly in *Science News:* "Sulfur Oxide Control: A Grim Future," August 29, 1970, pp. 187–188; and "Secrecy and Elitism in Science and Government," July 31, 1971, pp. 82–83.

37. Adrian J. B. Wood, J. Serge Taylor, Frederick R. Anderson, and Laurence I. Moss, *Strategies for Pollution Abatement*, draft of January 22, 1971, pp. S-2, 3-41.

38. Quoted in John F. Burby, "Environment Report/White House Activists Debate Form of Sulfur Tax; Industry Shuns Both," *National Journal*, vol. 4, no. 43, October 21, 1972, p. 1643.

39. Supporters of the tax approach are listed in *ibid.* Industry opposition is described in John F. Burby, "Environment Report/White House Plans Push for Sulfur Tax Despite Strong Industry Opposition," *National Journal*, vol. 4, no. 44, October 28, 1972.

40. National Academy of Sciences, *The Science Committee*, Washington, D.C., 1972, Appendix D, p. 46, Table D-6.

41. The circumstances are described in William H. Rodgers, Jr., *Corporate Country*, Rodale Press Inc., Emmaus, Pennsylvania, 1973, pp. 180–186.

42. National Academy of Sciences, *Fluorides*, Washington, D.C., 1971, p. 236; Center for Science in the Public Interest, *Fluorides in the Air*, Washington, D.C., 1972, p. 34.

43. National Research Council. Committee on Food Standards and Fortification Policy, *Nutritional Guidelines, Progress Report, August 28–November 30, 1970*, p. 14.

44. National Research Council, Committee on Food Standards and Fortification Policy, *Nutritional Guideline Recommendations—Frozen Convenience Dinners*, 1971, pp. 9, 11.

45. *Ibid.*, Table 6.

46. Council on Children, Media, and Merchandising, "Comments on FDA's

Proposed Nutritional Quality Guidelines," Submitted by Annette Dickinson to FDA Commissioner Charles Edwards, February 18, 1972.

47. Anita Johnson and Mary Goodwin, of Health Research Group, to HEW Hearing Clerk, February 25, 1972.

48. National Academy of Sciences, Highway Research Board, *National Study of the Composition of Roadside Litter*, A Report from the Highway Research Board to Keep America Beautiful, Inc., September 12, 1969, p. 2.

49. See, for example, Keep America Beautiful's newsletter, *Idea Service*, March 1968, p. 1.

50. Special thanks are accorded to the three KAB representatives on the acknowledgments page of the *Roadside Litter* study.

51. *Roadside Litter*, p. 9.

52. *Ibid.*, p. 2.

53. For a detailed review of the battle, see William H. Rodgers, Jr., "Ecology Denied: The Unmaking of a Majority," *Washington Monthly*, February 1971, pp. 39–43.

54. See, for example, the flyer distributed by a Citizens Committee against Initiative 256, reprinted in U.S. Congress. Senate, Committee on Commerce, *Solid Waste Management Act of 1972*. Hearings before the Subcommittee on the Environment, 92nd Cong., 2nd Sess., March 1972, p. 281.

55. *Ibid.*, pp. 271, 266.

56. *Ibid.*, p. 411.

57. Eileen Claussen, *Oregon's Bottle Bill: The First Six Months*, Environmental Protection Agency Publication SW-109, 1973, p. 5.

58. James L. Goddard, "The Drug Establishment," *Esquire*, March 1969. First published in *Esquire* magazine.

59. *Highway Research Board Directory*, 1970.

60. National Academy of Sciences, *Science and Technology in Presidential Policymaking: A Proposal*, Washington, D.C., June 1974, p. 6.

61. *Letter to Members*, vol. 1, no. 2, January 1970, p. 16.

62. Philip H. Abelson, "The New Physics Report," *Science*, vol. 177, no. 4048, August 11, 1972, p. 479. Copyright © 1972 by the American Association for the Advancement of Science.

63. *Letter to Members*, vol. 2, no. 6, p. 10.

64. National Academy of Sciences–National Research Council. *Chemistry: Opportunities and Needs*, Washington, D.C., 1965, pp. 20, 21.

65. *Ibid.*, p. 150.

66. "Chemists Out of Work," *Nature*, vol. 231, May 14, 1971, p. 78.

67. James A. Davis, et al., *Stipends and Spouses*, The University of Chicago

Press, Chicago, 1962, pp. 99, 127, 73, 59. This is the version that was ultimately published; it may differ somewhat from the version rejected by the Academy.

68. M. H. Trytten to Peter H. Rossi, August 22, 1961.

69. National Academy of Sciences, "On Potential Sources of Bias," August 20, 1971.

70. *Ibid.*, January 1973.

71. Joseph Gilbert, SAE secretary and general manager, "What Is Bias?" *Automotive Engineering*, April 1972.

CHAPTER FIVE

1. Informative discussions of the waste management problem can be found in the following documents: *Observations Concerning the Management of High-Level Radioactive Waste Material*, Report to the Joint Committee on Atomic Energy, United States Congress, by the Comptroller General of the United States, May 29, 1968, 48 pp., B-164052 (originally "secret" but declassified December 18, 1970); *Progress and Problems in Programs for Managing High-Level Radioactive Wastes*, Report to the Joint Committee on Atomic Energy by the Comptroller General of the United States, January 29, 1971, 106 pp., B-164052; United States Atomic Energy Commission, Division of Waste Management and Transportation, *Plan for the Management of AEC-Generated Radioactive Wastes*, WASH-1202, U.S. Government Printing Office, Washington, D.C., January 1972, 40 pp.; Charles H. Fox, *Radioactive Wastes*, published by U.S. Atomic Energy Commission, Division of Technical Information, Understanding the Atom Series, 1966 (rev. 1969), 46 pp.

2. Alvin M. Weinberg, "Social Institutions and Nuclear Energy," lecture presented at the annual meeting of the American Association for the Advancement of Science, Philadelphia, December 27, 1971, reprinted in *Science*, vol. 177, no. 4043, July 7, 1972, pp. 28, 32.

3. National Academy of Sciences, *The Disposal of Radioactive Waste on Land*, NAS-NRC Publ. 519, 1957, pp. 3, 1.

4. The advantages of salt are discussed in a speech by Frank K. Pittman, director of the AEC's Division of Waste Management and Transportation, "Long-Term Storage of High-Level Radioactive Wastes," presented at the Kansas Engineering Society Meeting, Topeka, Kansas, January 19, 1972.

5. W. C. McLain and R. L. Bradshaw, "Status of Investigations of Salt Formations for Disposal of Highly Radioactive Power-Reactor Wastes," *Nuclear Safety* (a bimonthly review prepared by the AEC and Oak Ridge National Laboratory), vol. 11, no. 2, March–April 1970, abstract.

6. H. H. Hess to John A. McCone, June 21, 1960.

7. A. R. Luedecke, AEC general manager, to H. H. Hess, January 4, 1961.

8. Atomic Energy Commission, "Comments on the Background of the May 1966 Report of the NAS Committee on Geologic Aspects of Radioactive Waste Disposal," March 1970, p. 4.

9. National Academy of Sciences–National Research Council, Committee on Geologic Aspects of Radioactive Waste Disposal, *Report to the Division of Reactor Development and Technology, United States Atomic Energy Commission*, May 1966, p. 1.

10. Earl Cook, "Benefit-Cost Decisions in Nuclear Energy Applications," testimony delivered at hearing on proposed nuclear power plant for Shoreham, Long Island, New York, February 1971.

11. Earl Cook to U.S. Senator Frank Church (D-Idaho), May 1, 1970.

12. *Report to the Division of Reactor Development and Technology, United States Atomic Energy Commission*, May 1966, pp. 11, 10, 42, 30, 70, 66, 76, 69.

13. Milton Shaw, director, Division of Reactor Development and Technology, to Frederick Seitz, November 7, 1966, plus attached "Comments on NAS Report."

14. Milton Shaw to Frederick Seitz, May 25, 1967.

15. Frederick Seitz to Milton Shaw, June 5, 1967.

16. Proposal for Committee on Radioactive Waste Disposal Advisory to the Atomic Energy Commission, with cover letter from Frederick Seitz to Glenn Seaborg, February 29, 1968.

17. The Church–Seaborg correspondence is reprinted in *Congressional Record*, 91st Cong., 2nd Sess., vol. 116, pt. 10, April 30, 1970, p. 13569.

18. Edmund S. Muskie to Glenn T. Seaborg, March 3, 1970, reprinted in U.S. Congress, Senate, Committee on Public Works, *Underground Uses of Nuclear Energy*, Hearings before the Subcommittee on Air and Water Pollution, 91st Cong., 1st Sess., November 1969, Appendix V, p. 461.

19. *New York Times*, March 7, 1970.

20. Atomic Energy Commission, "Comments on the Background of the May 1966 Report of the NAS Committee on Geologic Aspects of Radioactive Waste Disposal," March 1970, reprinted in *Underground Uses of Nuclear Energy*, p. 513.

21. National Academy of Sciences–National Research Council, *Radioactive Waste Management: An Interim Report of the Committee on Radioactive Waste Management*," February 17, 1970, p. 8. Reprinted in *ibid.*, p. 518.

22. John Lear, "Radioactive Ashes in the Kansas Salt Cellar," *Saturday Review*, February 19, 1972, reprinted in *Congressional Record*, vol. 118, pt. 4, February 16, 1972, p. 4130 (also AEC press release dated June 17, 1970).

23. National Academy of Sciences–National Research Council, *Disposal of*

Solid Radioactive Wastes in Bedded Salt Deposits, Report by the Committee on Radioactive Waste Management, U.S. Government Printing Office, Washington, D.C., November 1970, pp. 1, 2.

24. *Ibid.*, p. 7.

25. U.S. Congress, Joint Committee on Atomic Energy, *Hearings on AEC Authorizing Legislation Fiscal Year 1972*, Part 3, 92nd Cong., 1st Sess., March 1971, p. 1501.

26. *Ibid.*, p. 1373.

27. AEC Staff Report on American Salt Company Operations, reprinted in *Congressional Record*, vol. 117, pt. 28, October 15, 1971, p. 36452.

28. John A. Erlewine, general manager, AEC, to Edward J. Bauser, executive director, Joint Committee on Atomic Energy, September 30, 1971; reprinted in *ibid.*

29. AEC Press Release No. P-20, dated January 21, 1972, plus attached Chapter VII from the report.

30. AEC Press Release No. P-143, May 18, 1972.

31. Frank K. Pittman, director of AEC's division of waste management and transportation, prepared statement presented to FY 1974 Authorization Hearings before the Joint Committee on Atomic Energy, March 22, 1973, p. 12.

32. *Congressional Record*, vol. 117, pt. 28, October 15, 1971, p. 36451.

33. NAS–NRC, *Report to the Division of Reactor Development and Technology*, pp. 42, 75, 78, 79.

34. U.S. Atomic Energy Commission, *Bedrock Waste Storage Exploration*, Draft Environmental Statement, WASH-1511, January 1972, pp. 15, 50, 51, 16, Appendix A. The draft statement was never issued as a final statement because funds for the shaft and exploratory tunnels were not included in fiscal year 1973 appropriations.

35. National Academy of Sciences, *An Evaluation of the Concept of Storing Radioactive Wastes in Bedrock below the Savannah River Plant Site*, Report by the Committee on Radioactive Waste Management, 1972, pp. 3, 4, 2.

36. Press release No. 856 from the Savannah River Operations Office of the U.S. Atomic Energy Commission, Aiken, South Carolina, November 17, 1972.

37. See note 1.

CHAPTER SIX

1. Details on the development of the SST program were gleaned from three sources: George Chatham, "The Supersonic Transport," in U.S. Congress, House Committee on Science and Astronautics, *Technical Information for Congress*, Report to the Subcommittee on Science, Research and Development, Prepared by the Science Policy Research Division, Congressional Research Service, Library of Congress, 92nd

Cong., 1st Sess., Washington, U.S. Government Printing Office, 1971, pp. 685–748; U.S. Federal Aviation Agency, Office of Supersonic Transport Development, "United States Supersonic Transport Program: Chronology, Brief History, Research Contract Summary," SST 65-10, July 1965; and R. L. Bisplinghoff, "The Supersonic Transport," *Scientific American*, vol. 210, no. 6 (June 1964), pp. 25–35.

2. Timing details are in John Lear, "The Era of Supersonic Morality," *Saturday Review*, June 6, 1964, pp. 49–50; and National Academy of Sciences–National Research Council; *U.S. Federal Aviation Agency Sonic Boom Test Program in Oklahoma City, Oklahoma*, report of a special advisory panel, Washington, D.C., June 8, 1964.

3. U.S. Federal Aviation Agency, Press Release 64-50, May 20, 1964.

4. Lear, "Supersonic Morality," p. 50.

5. Article filed to *Oklahoma City Times* from its Washington bureau, May 20, 1964.

6. Article filed to *The Daily Oklahoman* from its Washington bureau, May 20, 1964.

7. *Tulsa Tribune*, May 20, 1964.

8. M. King Hubbert to William A. Shurcliff and John T. Edsall, October 10, 1968.

9. National Academy of Sciences, *Test Program in Oklahoma City*, p. 7ff.

10. Contract No. FA-WA-65-4, dated September 1, 1964, between Federal Aviation Agency and National Academy of Sciences, with eleven subsequent modifications.

11. National Academy of Sciences, *Status Report, Committee on SST–Sonic Boom* (Washington, D.C., January 27, 1965), p. 16.

12. National Academy of Sciences, *Status Report, Committee on SST–Sonic Boom*, Washington, D.C., July 21, 1965, p. 23.

13. *Ibid.* (list of staff members).

14. National Academy, *Status Report*, January 27, 1965, pp. 28–29.

15. National Academy of Sciences, *Report on Generation and Propagation of Sonic Boom*, Washington, D.C., October 1967, p. 2.

16. National Academy of Sciences, *1968 Progress Report, Sonic Boom Generation and Propagation*, Washington, D.C., June 1968, p. 6.

17. National Academy of Sciences, *Report on Physical Effects of the Sonic Boom*, Washington, D.C., February 1968, p. 12.

18. William A. Shurcliff to John R. Dunning, May 29, 1968.

19. John C. Calhoun, Jr., et al., *Report to the Secretary of the Interior of the Special Study Group on Noise and Sonic Boom in Relation to Man*, Washington, D.C., U.S. Department of the Interior, November 1968, mimeographed, p. 5.

20. William A. Shurcliff and John T. Edsall to members of the National Academy of Sciences, August 8, 1968.

21. John S. Coleman to the Members of the Academy, August 22, 1968.

22. "Statement of the Committee on SST–Sonic Boom," August 19, 1968.

23. Excerpts from petitions in the files of William A. Shurcliff.

24. National Academy of Sciences, *News Report*, vol. 19, no. 2 (February 1969), p. 11.

25. Ralph Tyler Smith to Jeffrey R. Short, Jr., May 28, 1970.

26. Philip Handler to William A. Shurcliff, June 23, 1970.

27. Philip Handler to Mary Riecken, Trade Books Division, Prentice-Hall, Inc., July 10, 1970.

28. National Academy of Sciences, press release prepared for release on March 5, 1968.

29. National Academy of Sciences, *Report on Human Response to the Sonic Boom*, Washington, D.C., June 1968, p. 5.

30. Raymond A. Bauer to William A. Shurcliff, July 15, 1968.

31. *Human Response to Sonic Boom*, pp. 2, 12.

32. William A. Shurcliff to Raymond A. Bauer, July 2, 1968.

33. National Academy, *Status Report*, January 27, 1965, pp. 2–3; *Status Report*, July 21, 1965, p. 1.

34. The White House, "Memorandum for President, National Academy of Sciences; Administrator, Federal Aviation Agency; Secretary of Commerce," May 20, 1964.

35. Raymond A. Bauer, "Some Thoughts on Human Response to Sonic Boom," an address to the American Institute of Aeronautics and Astronautics, Philadelphia, Pa., October 22, 1968, p. 9.

36. U.S. Congress, House Committee on Appropriations, *Department of Transportation and Related Agencies Appropriations for 1970—Part 3 (Civil Supersonic Aircraft Development)* Hearings before a Subcommittee, 91st Cong., 1st Sess., October 9, 1969, p. 314.

37. National Academy of Sciences, *Weather and Climate Modification, Problems and Prospects* (Washington, D.C., 1966), vol. 1, p. 11.

38. James E. McDonald, "Statement Submitted for the Record at Hearings before the House Subcommittee on Transportation Appropriations," March 2, 1971, reprinted in U.S. Congress, Senate *Congressional Record*, vol. 117, pt. 6, March 19, 1971, pp. 7252–7255.

39. U.S. Congress, Senate *Congressional Record*, vol. 116, pt. 32, December 21, 1970, p. 43040.

40. U.S. Congress, House *Congressional Record*, 92nd Cong., 1st Sess., March 9, 1971, p. 5631.

41. U.S. Congress, Senate *Congressional Record*, vol. 117, pt. 6, March 19, 1971, p. 7264.

42. *Washington Post*, June 17, 1971.

43. *Huntsville* (Ala.) *Times*, March 3, 1971.

44. *Washington Post*, July 13, 1972.

45. *New York Times*, May 30, 1971.

46. National Academy of Sciences–National Research Council, *Summary Statement of the Ad Hoc Panel on (NOx) and the Ozone Layer* (Washington, D.C., July 29, 1971), p. 3.

47. National Academy of Sciences–National Academy of Engineering, *Biological Impacts of Increased Intensities of Solar Ultraviolet Radiation*, Washington, D.C., 1973, pp. 30, 31. This report stemmed from a November 1971 three-day meeting but was long delayed in publication.

48. *Ibid.*, p. 8.

49. Alan J. Grobecker, "Progress Report on the Climatic Impact Assessment Program," presented at the Third Conference on the Climatic Impact Assessment Program, Cambridge, Mass., February 25–March 1, 1974.

50. *Report of the SST Ad Hoc Review Committee*, reprinted in U.S. Congress, House *Congressional Record*, vol. 115, pt. 124, October 31, 1969, p. 32607.

51. Lee A. DuBridge, *ibid.*, p. 32609.

52. Richard L. Garwin, et al., *Final Report of the Ad Hoc Supersonic Transport Review Committee*, March 30, 1969, p. 10.

53. *Ibid.*, p. 6.

54. Lee A. DuBridge to Senator William Proxmire, April 22, 1970, reprinted in U.S. Congress, Senate *Economic Analysis and the Efficiency of Government, Part 4—Supersonic Transport Development*, Hearings before the Subcommittee on Economy in Government, Joint Economic Committee, 91st Cong., 2d Sess., May 1970, pp. 1029–1030.

55. *Science and Government Report*, vol. 1, no. 1, February 1, 1971, p. 2.

CHAPTER SEVEN

1. Information compiled from U.S. Congress, House Committee on Science and Astronautics, *A Technology Assessment of the Vietnam Defoliation Matter: A Case History*, Report to the Subcommittee on Science, Research and Development, prepared by Franklin P. Huddle of the Science Policy Research Division, Legislative Reference Service, Library of Congress, 91st Cong., 1st Sess., Washington, U.S. Government Printing Office, 1969, 73 pp.; W. B. House, et al., *Assessment of Ecological Effects of Extensive or Repeated Use of Herbicides*, Midwest Research Institute Project No. 3103-B, Final Report, August 15–December 1, 1967, 369 pp. (hereinafter called "MRI Report"), sponsored by Advanced Research Projects Agency of Department of Defense; American Association for the Advancement of Science, Herbicide Assessment Commission, *Background Material Relevant to Presentations at the 1970 Annual Meeting of the AAAS* (preliminary unpublished paper), 42 pp.; Philip M. Boffey, "Herbicides in Vietnam: AAAS study Finds Widespread Devastation," *Science*, vol. 171 (Janu-

ary 8, 1971), pp. 43–47; National Academy of Sciences, *The Effects of Herbicides in South Vietnam, Part A*, Washington, D.C., 1974.

2. Huddle, *Technology Assessment*, p. 13.

3. American Association for the Advancement of Science, "Chronological Summary of AAAS Actions Related to Proposals Concerning the War in Vietnam," prepared by Dael Wolfle, December 12, 1968 (unpublished), pp. 1, 2.

4. AAAS, Agenda for meeting of board of directors, October 21–22, 1967, Tab D.

5. Elinor Langer, "Chemical and Biological Warfare (I): The Research Program," *Science*, vol. 155 (January 13, 1967), p. 178.

6. National Academy of Sciences, *Report for 1944–45*, p. 7; *Report for 1945–46*, p. 2; Report for 1917, pp. 52, 53, 61.

7. Gale E. Peterson, "The Discovery and Development of 2,4-D," *Agricultural History*, vol. 41, no. 3 (July 1967), pp. 246–247.

8. National Academy of Sciences, *Report for 1946–47*, p. 4.

9. Cited in a memorandum prepared for the study team by the Academy's Office of the Archivist, July 25, 1973.

10. Langer, "Chemical and Biological Warfare: I," pp. 176, 178.

11. Don K. Price to Robert S. McNamara, September 13, 1967, reproduced in Agenda for October 21–22, 1967, meeting of AAAS board of directors, Tab E.

12. For the sequence of events, see Huddle, *Technology Assessment*, p. 35, n. 71.

13. John S. Foster, Jr., to Don K. Price, September 29, 1967, reproduced in Agenda for October 21–22, 1967, meeting of AAAS board of directors, Tab E.

14. See work statement reproduced in Huddle, *Technology Assessment*, p. 35.

15. Frederick Seitz to John S. Foster, January 31, 1968, released as cover letter to the Academy's review of the MRI Report.

16. Peterson, "Discovery and Development of 2,4-D," p. 247.

17. Details of the Academy–MRI interaction are from Huddle, *Technology Assessment*, p. 38; and from the NAS review of the MRI Report.

18. Memorandum from A. G. Norman to Frederick Seitz, January 29, 1968.

19. Frederick Seitz to John S. Foster, Jr., January 31, 1968.

20. Huddle, *Technology Assessment*, p. 38.

21. *New York Times*, February 13, 1968.

22. Huddle, *Technology Assessment*, p. 41.

23. "Book Reviews," *Ecology*, vol. 49, no. 6 (Autumn 1968), p. 1212.

24. *Ibid.*, pp. 1214–1215.

25. *Ibid.*, p. 1215.

26. Sheldon Novick, "The Vietnam Herbicide Experiment," *Scientist and Citizen*, vol. 10, no. 1 (January–February 1968), pp. 20–21.

27. AAAS, "Agenda for Meeting of Board of Directors," June 15–16, 1968, Tab B.

28. "On the Use of Herbicides in Vietnam: A Statement by the Board of Directors of the American Association for the Advancement of Science," *Science*, vol. 161 (July 19, 1968), p. 254.

29. "Meselson to Head Herbicide Study," *Science*, vol. 167 (January 2, 1970), p. 37.

30. Meselson's task is defined in the minutes of the meeting of the AAAS board of directors, December 28–29, 1969, Agenda Item 23; and in a letter from Meselson to Dael Wolfle, then–executive officer of the AAAS, dated January 29, 1970, which was agreed to by Bentley Glass, then-president of the AAAS, in a letter to Meselson dated February 9, 1970.

31. Philip M. Boffey, "Herbicides in Vietnam: AAAS Study Runs into a Military Roadblock," *Science*, vol. 170 (October 2, 1970), p. 42.

32. John Constable and Matthew Meselson, "The Ecological Impact of Large-Scale Defoliation in Vietnam," *Sierra Club Bulletin*, April 1971, reprinted in U.S. Congress, House *Congressional Record*, vol. 117, pt. 13, June 2, 1971, pp. 17692–17694.

33. AAAS Herbicide Assessment Commission, *Summary of Presentations* given on December 29, 1970, accompanied by *Background Material Relevant to Presentations at the 1970 Annual Meeting of the AAAS*; reprinted in *Congressional Record*, vol. 118, March 3, 1972, pp. 6806–6813.

34. The White House, Press release dated December 26, 1971.

35. U.S. Congress, Senate *H.R. 17123* (the authorization act for military procurement), 91st Cong., 2nd Sess., Section 506(c).

36. U.S. Congress, Senate *Congressional Record*, vol. 116, pt. 22, August 25, 1970, p. 30007.

37. U.S. Congress, Senate *Congressional Record*, vol. 116, pt. 22, August 26, 1970, p. 30052.

38. U.S. Department of the Army. Office. Chief of Engineers. Engineer Strategic Studies Group, *Herbicides and Military Operations*, February 1972; vols. I and II are unclassified; vol. III, containing the contingency plans, is secret, but its contents are summarized in *Science & Government Report*, vol. 2, no. 11, August 18, 1972, p. 1.

39. The difficulties are described in National Academy of Sciences, *The Effects of Herbicides in South Vietnam*, Part A, Washington, D.C., 1974, pp. I-4, I-5, I-10, I-11.

40. The struggle between reviewers and the committee is described in Deborah Shapley, "Herbicides: Academy Finds Damage in Vietnam

after a Fight of Its Own," *Science*, vol. 183, no. 4130, March 22, 1974, p. 1177; and in "Herbicide Report Leaked to Avert DOD Mishandling," *Science & Government Report*, vol. 4, no. 5, March 1, 1974, p. 5.

41. National Academy, *Effects of Herbicides*, p. S-9.

42. *Ibid.*, S-8, xiii–xv, statements of exception.

43. *Ibid.*, VII-64, VII-65, VII-66; VII-1, VII-2; x; VII-3, VII-4; S-11; VII-13; xi.

44. See, for example, *New York Times*, February 22, 1974. The National News Council subsequently criticized the *Times* for its handling of the story.

45. Press release No. 74-32 from Senator Gaylord Nelson.

46. Department of Defense Comments, reprinted in *Congressional Record*, vol. 120, no. 24, February 28, 1974, p. S2427. Meselson et al. replied in *Science*, November 15, 1974, p. 584.

47. Kenneth V. Thimann, "Herbicides in Vietnam," a letter to the editor, *Science*, July 19, 1974, p. 207.

48. D. S. Greenberg, "Defoliation: AAAS Study Delayed by Resignations from Committee," *Science*, vol. 159 (February 22, 1968), pp. 857–859.

49. Roy M. Sachs, "Vietnam: AAAS Herbicide Study," *Science*, vol. 170 (December 4, 1970), pp. 1034–1036.

CHAPTER EIGHT

1. National Academy of Sciences–National Research Council, *Food Protection Committee*, pamphlet dated January 1971, p. 6.

2. Julius M. Coon, "Protecting Our Internal Environment," *Nutrition Today*, vol. 5, no. 2 (Summer 1970) pp. 14ff. Reprinted with permission of *Nutrition Today*. Copyright © 1970 by Nutrition Today, Inc.

3. L. Golberg, "Trace Chemical Contaminants in Food: Potential for Harm," paper presented at symposium sponsored by the Food and Drug Directorate, Department of National Health and Welfare, Ottawa, Ontario, June 1970; printed in *Fd. Cosmet. Toxicol.*, vol. 9, pp. 65–80, Pergamon Press 1971; reprinted in "Chemicals and the Future of Man," Hearings before the Subcommittee on Executive Reorganization and Government Research, Committee on Government Operations, U.S. Senate, 92nd Cong., 1st Sess., April 1971, Exhibit 12.

4. National Academy of Sciences–National Research Council, *Guidelines for Estimating Toxicologically Insignificant Levels of Chemicals in Food*, 1969, p. 1.

5. *Ibid.*, p. 3.

6. *Ibid.*, p. 6.

7. *Ibid.*, p. 7.

8. 21 CFR 121.6 (Rev. January 1, 1972).

9. Ad Hoc Committee on the Evaluation of Low Levels of Environmental Chemical Carcinogens, *Evaluation of Environmental Carcinogens*,

Report to the Surgeon General, USPHS, April 22, 1970, reprinted by U.S. Department of Health, Education and Welfare, National Institutes of Health, pp. 1, 2, 4, 8.

10. *Ibid.*, pp. 1, 7, Appendix II.

11. Philip Handler to Jesse Steinfeld, December 14, 1970.

12. Summary of Meeting on Interpretation of "Guidelines for Estimating Toxicologically Insignificant Levels of Chemicals in Food," undated and unsigned.

13. Philip Handler to Representative L. H. Fountain, chairman of the House Intergovernmental Relations Subcommittee, April 10, 1972, and September 6, 1972. Reprinted in U.S. Congress, House of Representatives, *Regulation of Diethylstilbestrol (DES), Part 3,* Hearings before a Subcommittee of the Committee on Government Operations, 92nd Cong., 2nd Sess., August 15, 1972, pp. 431, 437.

14. U.S. Congress, House Committee on Government Operations, *Regulation of Cyclamate Sweeteners,* House Report No. 91-1585, 91st Cong., 2nd Sess., Washington, U.S. Government Printing Office, 1970, p. 3.

15. Herbert L. Ley, Jr., Commissioner of Food and Drugs, statement at press conference, Washington, D.C., April 3, 1969.

16. See note 14.

17. *Regulation of Cyclamate Sweeteners,* p. 12.

18. National Academy of Sciences–National Research Council, Food and Nutrition Board, *Policy Statement on Artificial Sweeteners,* adopted November 1954, published November 1955, pp. 7–8.

19. National Academy of Sciences–National Research Council, Food Protection Committee, *The Safety of Artificial Sweeteners for Use in Foods,* NAS–NRC Pub. No. 386, August 1955, pp. 7–8.

20. *Policy Statement,* p. 5.

21. Frederick Stare, Harvard nutritionist, quoted in *New Republic,* January 4, 1969.

22. *Policy Statement,* p. 8.

23. *Policy Statement,* Revised April 1962, p. 8.

24. Quoted in *Wall Street Journal,* September 9, 1964; *New York Times,* September 10, 1964.

25. National Academy of Sciences–National Research Council, Food Protection Committee, *Nonnutritive Sweeteners,* Interim Report to the U.S. Food and Drug Administration, November 1968, pp. 2, 8, 9.

26. FDA Press Release 69-21, April 3, 1969.

27. *Review and Recommendations on the NAS/NRC Food Protection Committee Report on the Non-Nutritive Sweeteners* (For Official Use Only), December 11, 1968, p. 1. Submitted with cover letter from John J. Schrogie and Herman F. Kraybill to Herbert L. Ley, Jr., Commissioner of Food and Drugs, December 12, 1968.

28. *Ibid.*, pp. 1, 2.

29. *Review and Recommendations* by Schrogie–Kraybill, p. 4.

30. The interim report is referred to in letter from William Summerson to Paul Johnson, April 18, 1968.

31. M. S. Legator, et al., "Cytogenetic Studies in Rats of Cyclohexylamine, a Metabolite of Cyclamate," *Science*, vol. 165, no. 3898, September 12, 1969, p. 1139.

32. Samuel S. Epstein, et al., "Wisdom of Cyclamate Ban," letter to the editor, *Science*, vol. 166, no. 3913, December 26, 1969, p. 1575.

33. U.S. Congress, House Committee on Government Operations, *Cyclamate Sweeteners*, subcommittee hearing, June 10, 1970, pp. 14–23.

34. See note 26.

35. *Regulation of Cyclamate Sweeteners*, pp. 5, 6.

36. *Cyclamate Sweeteners* hearing, p. 24.

37. Jack Schubert, speech to Federal Bar Association's Food and Drug Law Committee, November 24, 1969, Washington, D.C., quoted in *Food Chemical News*, December 1, 1969, pp. 8–10.

38. The sequence of events is described in "Summary of Scientific Events Regarding Food & Drug Research Laboratories, Inc., Data during the 10 Days Preceding Secretary Finch's Order on Cyclamate Sweeteners," a four-page press release by Abbott Laboratories, October 29, 1969.

39. *Statement of the Committee on Nonnutritive Sweeteners*, October 17, 1969.

40. Philip Handler to Herbert L. Ley, Jr., October 24, 1969.

41. Jesse L. Steinfeld, statement at press conference, October 18, 1969.

42. National Academy of Sciences–National Research Council, *Investigation of Teratogenic and Mutagenic Effects of Chemicals Intended for Use in Foods*, October 30, 1969, pp. 3, 5.

43. Philip Handler to Herbert Ley, October 31, 1969.

44. James F. Crow to Paul Johnson, October 30, 1969.

45. "Sombre Greeting from Abroad," *Nature*, vol. 224, December 27, 1969, p. 1250.

46. Philip Handler, "Science and the Federal Government: Which Way to Go?" speech to Federation of American Societies for Experimental Biology, Atlantic City, N.J., April 14, 1971, p. 7.

47. Philip Handler to Samuel S. Epstein, January 15, 1970.

48. *Report of the Medical Advisory Group on Cyclamates*, August 1970, Reprinted in U.S. Congress, House Committee on Government Operations, *The Safety and Effectiveness of New Drugs (Market Withdrawal of Drugs Containing Cyclamates)*, Hearing, 92nd Cong., 1st Sess., May 3, 1971, pp. 20–22.

49. Herbert H. Schaumburg et al., "Monosodium L-Glutamate: Its Phar-

macology and Role in the Chinese Restaurant Syndrome," *Science*, vol. 163, February 21, 1969, pp. 826–828.

50. J. W. Olney, *J. Neuropathol. Exp. Neurol.* 28, 455 (1966); J. W. Olney, *Science*, 164, 719 (1969). See also U.S. Congress, Senate Select Committee on Nutrition and Human Needs, *Nutrition and Human Needs, Part 13A—Nutrition and Private Industry.* Hearings 91st Cong., 1st Sess., July 1969, pp. 4014–4020.

51. *Washington Post*, October 24, 1969.

52. *Nutrition and Human Needs, Part 13A*, p. 3910.

53. National Academy of Sciences–National Research Council, Food Protection Committee, *Safety and Suitability of Monosodium Glutamate for Use in Baby Foods*, July 1970, p. 37.

54. Arthur T. Schramm to William J. Darby, October 30, 1969.

55. National Academy, *Safety and Suitability of Monosodium Glutamate*, p. 42, ref. 10. Filer was associated with Stegink on this study.

56. *Ibid.*, p. 21, ref. 3.

57. George M. Owen, "Modification of Cow's Milk for Infant Formulas: Current Practices," *The American Journal of Clinical Nutrition*, vol. 22, no. 8, August 1969, pp. 1150–1155.

58. "Staff Comments on Allegations Made by Witnesses before the Senate Select Committee," Attachment Seven to letter from Philip Handler to Senator Gaylord Nelson, October 13, 1972; reprinted in hearings on *Nutrition and Human Needs—1972, Part 4C—Food Additives*, p. 1730.

59. *Safety and Suitability of Monosodium Glutamate*, p. 2.

60. *Nutrition and Human Needs—1972, Part 4A—Food Additives*, p. 824.

61. *Nutrition and Human Needs—1972, Part 4C—Food Additives*, p. 1729.

62. John W. Olney to Philip Boffey, June 6, 1972.

63. Olney's charges are in *Nutrition and Human Needs—1972, Part 4A—Food Additives*, pp. 824–829. He does not specifically name Oser and Coulston but refers to their research teams as groups A and B. Coulston's response, in a letter to Senator Charles Percy (R-Ill.), is quoted in *Food Chemical News*, October 9, 1972, p. 18.

64. Quoted in "What Ever Happened to Baby Food?" an unpublished report to Ralph Nader by James S. Turner.

65. New York *Journal of Commerce*, April 29, 1971.

66. *Food Chemical News*, vol. 13, no. 47 (February 14, 1972), p. 15.

67. *Nutrition and Human Needs, Part 13A*, p. 4027.

68. *Food Chemical News*, July 5, 1971, p. 34.

69. Jean Carper, "Food Coloring," *The Wednesday News and Opinion* (an experimental paper published by the Morning News Company, Washington, D.C.), May 24, 1972, p. 16.

70. *Ibid.*

71. Both Russian studies are discussed in the National Academy of Sciences–National Research Council, Committee on Food Protection, *Report of Ad Hoc Subcommittee on the Evaluation of Red No. 2*, June 1972, p. 3.

72. The FDA's sentiments are described in *Ibid.*, p. 3; a similar view is expressed by Michael F. Jacobson in *Food Colors*, a booklet published by Center for Science in the Public Interest, Washington, D.C., March 13, 1972.

73. The FDA studies are cited in *Report of the Ad Hoc Subcommittee*, pp. 4–5. The reinterpretation of the Stanford Research Study is cited in a letter from J. M. Coon to Philip Handler, November 27, 1972.

74. *Federal Register*, September 11, 1971 (36 F.R. 18336).

75. Memorandum of meeting of Bureau of Foods scientists, November 18, 1971.

76. *Food Chemical News*, December 6, 1971, p. 26.

77. Philip Handler to Anita Johnson and Sidney Wolfe, March 8, 1972.

78. *Report of the Ad Hoc Subcommittee*, June 1972, p. 8.

79. Philip Handler to Charles C. Edwards, June 13, 1972.

80. *Nutrition and Human Needs—1972, Part 4C—Food Additives*, p. 1724.

81. HEW News press release 72-63, July 3, 1972; *Federal Register*, July 4, 1972 (21 CFR Part 8).

CHAPTER NINE

1. Rachel Carson, speech to Women's National Press Club, Washington, D.C., December 5, 1962.

2. *Ibid.*

3. The President's Science Advisory Committee, *Use of Pesticides*, Washington, D.C., U.S. Government Printing Office, May 15, 1963, p. 23.

4. Quoted in Frank Graham, Jr., *Since Silent Spring* (Boston, Houghton Mifflin Company, 1970), p. 39.

5. Quoted in *ibid.*, p. 88.

6. I. L. Baldwin, "Chemicals and Pests," *Science*, vol. 137 (September 28, 1962), pp. 1042–1043.

7. William J. Darby, "Silence, Miss Carson," *Chemical and Engineering News*, vol. 40, no. 40 (October 1, 1962), pp. 60–63.

8. See Graham, *Since Silent Spring*, p. 59.

9. Quoted in "Miss Carson's Critics Are Criticized," *Medical Economics*, November 18, 1963, p. 334.

10. Baldwin, see note 6.

11. Quoted in Graham, *Since Silent Spring*, p. 39.

12. National Academy of Sciences–National Research Council, *Part I: Evaluation of Pesticide–Wildlife Problems*, Publication 920-A, Washington, D.C., 1962, p. 10.

13. National Academy of Sciences–National Research Council, *Part II: Policy and Procedures for Pest Control*, Publication 920-B, Washington, D.C., 1962, p. 1.

14. Roland C. Clement, "Pest Control and Wildlife Relationships," *Audubon*, vol. 64, no. 6, November–December 1962, p. 358. Reprinted from *Audubon*, the magazine of the National Audubon Society, Copyright © 1962.

15. Frank E. Egler, "Pesticides and the National Academy of Sciences," *Atlantic Naturalist*, October–December 1962, pp. 269, 270.

16. C. M. Tarzwell to Ira Gabrielson, October 29, 1962.

17. Clarence Cottam to Ira Gabrielson, November 2, 1962.

18. Cottam to Philip M. Boffey, May 19, 1971.

19. National Academy of Sciences, *Land Use and Wildlife Resources*, Washington, D.C., 1970, pp. 186–188.

20. PSAC, *Use of Pesticides*, p. 20.

21. National Academy of Sciences–National Research Council, Pesticide Residues Committee, *Report on "No Residue" and "Zero Tolerance,"* Washington, D.C., June 1965, p. 16.

22. NAS, *Zero Tolerance Report*, p. 7.

23. See Statement for Implementation of the NRC Pesticide Residue Committee's *Report on "No Residue" and "Zero Tolerance,"* Federal Register, vol. 31, no. 71 (April 13, 1966), pp. 5723–5724.

24. NAS, *Zero Tolerance Report*, p. 17.

25. PSAC, *Use of Pesticides*, pp. 17–18, U.S. Department of Health, Education, and Welfare; *Report of the Secretary's Commission on Pesticides and Their Relationship to Human Health* (the Mrak Report), U.S. Government Printing Office, Washington, D.C., December 1969, p. 7.

26. Comptroller General of the United States, *Need to Improve Regulatory Enforcement Procedures Involving Pesticides*, Report to the Congress, No. B-133192, September 10, 1968, pp. 14, 20.

27. U.S. Congress, House Committee on Government Operations, *Deficiencies in Administration of Federal Insecticide, Fungicide, and Rodenticide Act*, House Report No. 91-637, 91st Cong., 1st Sess., 1969, pp. 13–16.

28. This is revealed in the minutes of a later committee, the Committee on Persistent Pesticides, September 30, 1968, p. 11.

29. Details on the development of the fire ant program, and of opposition to it, can be found in Rachel Carson, *Silent Spring*, Boston, Houghton Mifflin Co., 1962, Chap. 10; Harrison Wellford, *Sowing the Wind: A Report from Ralph Nader's Center for Study of Responsive Law on*

Food Safety and the Chemical Harvest, New York, Grossman Publishers, 1972, Chap. 11; and an unpublished paper by John Blodgett.

30. The Comptroller General of the United States, *Weaknesses and Problem Areas in the Administration of the Imported Fire Ant Eradication Program,* Report to the Congress, B-133192, January 1965.

31. Excerpts from the 1958 report are included in a memorandum from Harlow B. Mills to Drs. Smith and Newsom of the Academy committee, August 29, 1967. The memorandum indicates that Mills did not circulate his previous report to the committee until after the committee had essentially completed its deliberations.

32. *Silent Spring,* pp. 163, 172.

33. National Academy of Sciences, *Report of the Committee on the Imported Fire Ant to Administrator, Agricultural Research Service, U.S. Department of Agriculture,* September 28, 1967, published in *Congressional Record,* vol. 117, pt. 19 July 15, 1971, pp. 25423–25425.

34. U.S. Congress, House Committee on Appropriations, *Hearings on Department of Agriculture Appropriations for 1969,* Part 2, p. 140.

35. Kenneth C. Walker, Assistant to Associate Administrator for Research, Agricultural Research Service, USDA, to John E. Blodgett, January 27, 1971.

36. Figures supplied by Animal and Plant Health Inspection Service, USDA, October 15, 1973.

37. For a fuller discussion of fire ant politics, see Wellford, *Sowing the Wind,* Chap. 11.

38. "Plan-of-Work," Contract No. 12-14-100-9468(86) between U.S. Department of Agriculture and National Academy of Sciences, June 15, 1967.

39. *Ibid.*

40. National Research Council, *Report of the Committee on Persistent Pesticides,* Washington, D.C., May 27, 1969, pp. 27–30.

41. *Ibid,* p. 30.

42. Memorandum from Harvey E. Sheppard to Members of the Committee on Persistent Pesticides, March 10, 1969.

43. "Government DDT Ban Could Be Permanent," *Industrial Research,* August 1969, p. 331.

44. "Pesticides Praised," *Chemical Week,* June 21, 1969, p. 55.

45. For the Washington and Nebraska hearings, see *DDT: Selected Statements from State of Washington DDT Hearings and Other Related Papers,* published by the DDT Producers of the United States and compiled by Max Sobelman, Montrose Chemical Corporation of California (May 1970), pp. 6, 254.

46. Charles F. Wurster to Senator Philip A. Hart, July 18, 1969.

47. "The Academy of Sciences Lays Another Thin-Shelled Egg," *Audubon,*

vol. 71, no. 104, July 1969, p. 103. Reprinted from *Audubon*, the magazine of the National Audubon Society. Copyright © 1969.

48. Material in this paragraph is drawn from U.S. Department of Health, Education, and Welfare, *Report of the Secretary's Commission on Pesticides and Their Relationship to Environmental Health, Parts I and II*, Washington, D.C., U.S. Government Printing Office, 1969, pp. xiii–xvii, 5, 7–10, 37, 470, 657–658.

49. The Mrak Commission figures are in a letter from W. Wade Talbot, executive officer, HEW Secretary's Pesticide Advisory Committee, to John E. Blodgett, September 3, 1970.

50. NRC, *Report of the Committee on Persistent Pesticides*, p. 3.

51. Committee on Persistent Pesticides, Minutes, August 1–3, 1968, p. 36.

52. Minutes, March 15–16, 1968, p. 2.

53. Minutes, March 1–3, 1968, p. 331.

54. Clement, "Thin-Shelled Egg," p. 103.

55. National Academy of Sciences, *Chlorinated Hydrocarbons in the Marine Environment*, Washington, D.C., 1971, pp. 1, 2.

56. "Exaggeration: The Other Pollution Peril," *Nation's Business*, April 1971, p. 32.

57. Philip Handler, "The Federal Government and the Scientific Community," *Science*, vol. 171 (January 15, 1971), p. 1481.

CHAPTER TEN

1. An "advance copy," entitled *Airborne Lead in Perspective*, was released in September 1971. A final version, entitled *Lead: Airborne Lead in Perspective*, was published in 1972.

2. Robert Gillette, "Lead in the Air: Industry Weight on Academy Panel Challenged," *Science*, vol. 174, no. 4011 (November 19, 1971), p. 802. Copyright © 1971 by the American Association for the Advancement of Science.

3. Gillette, "Lead in the Air," p. 801.

4. Tsaihwa J. Chow to Charles Malone, NRC, August 29, 1971.

5. Louise H. Marshall to Tsaihwa J. Chow, September 14, 1971.

6. Gillette, "Lead in the Air," p. 801.

7. *Ibid.*

8. National Academy of Sciences–National Research Council, *Lead: Airborne Lead in Perspective*, Washington, D.C., 1972, p. viii.

9. *Ibid.*, pp. 205–206.

10. *Ibid.*, p. 206.

11. *Los Angeles Times*, October 10, 1971.

12. National Research Council, press release for morning papers of Tuesday, September 7, 1971.

13. *Lead: Airborne Lead in Perspective,* p. 209.

14. *Ibid.*

15. *Ibid.,* p. 210.

16. AP story carried in *Detroit Free-Press,* September 7, 1971.

17. *Chemical and Engineering News,* September 13, 1971.

18. *Sioux Falls* (S.C.) *Argus-Leader,* September 19, 1971.

19. *Wall Street Journal,* September 8, 1971.

20. *Barron's,* September 27, 1971.

21. Ethyl press release CPR91471, dated September 14, 1971.

22. See, for example, John L. Kimberley, executive vice-president, Lead Industries Association, Inc., remarks delivered at press conference, Washington, D.C., November 17, 1971.

23. *Los Angeles Times,* October 10, 1971.

24. *Ibid.*

25. "Lead Balloon," *Newsweek,* page 74, December 6, 1971.

26. Environmental Protection Agency, press release for June 4, 1971, with attached interim report of Seven-City Study.

27. *Airborne Lead in Perspective,* p. 20 of the advance copy. This sentence was dropped from the final printed version.

28. *Ibid.,* p. 312 of advance copy, p. 206 of final version.

29. *Ibid.,* final version, pp. 25, 206.

30. State of Illinois, Institute for Environmental Quality, *A Study of Environmental Pollution by Lead,* IIEQ Document No. 71-7, November 1971, pp. 129, 125, 5–6, abstract, 126.

31. Julian M. Sturtevant, letter to the editor of *Chemical and Engineering News,* October 18, 1971.

32. *EDF Newsletter,* June 1970.

33. Joshua Lederberg, "Leaded Gasoline Presents Hazards to Health," *Washington Post,* February 22, 1969.

34. D. Bryce-Smith to Philip M. Boffey, January 31, 1972.

35. Environmental Protection Agency press release dated February 23, 1972, supported by backup document entitled "Health Hazards of Lead."

36. P. B. Hammond to Orin W. Stopinski of EPA's National Environmental Research Center, January 17, 1972; P. B. Hammond to Douglas J. Hammer, of EPA's Division of Health Effects Research, February 28, 1972; P. B. Hammond to Vaun Newill, assistant for health effects in EPA's Office of Research and Monitoring, June 8, 1972.

37. Illinois Institute for Environmental Quality, *Environmental Lead: Health Effects and Recommended Standards*, November 1972, p. 1.

38. Environmental Protection Agency, "Regulation of Fuels and Fuel Additives," *Federal Register*, vol. 38, no. 234, Part III, December 6, 1973, p. 33734.

CHAPTER ELEVEN

1. Harvey Brooks to George Wald, December 8, 1969.

2. National Academy of Sciences, *Rehabilitation Potential of Western Coal Lands*, Ballinger Publishing Company, Cambridge, Mass., 1974 (limited number of draft copies distributed in 1973). The report and its impact on Congress are discussed in two articles by Robert Gillette, "Strip-Mining: House, Senate Gird for Renewed Debate," and "Western Coal: Does the Debate Follow Irreversible Commitment?" *Science*, vol. 181, August 10, 1973, p. 524ff., and vol. 182, November 2, 1973, p. 456ff.

3. National Academy of Sciences, *Natural Resources*, Washington, D.C., 1962, p. 12. The accuracy of the prediction, which had also been made earlier outside the Academy context by the same author, is discussed in Robert Gillette, "Oil and Gas Resources: Did the USGS Gush Too High?" *Science*, vol. 185, July 12, 1974, p. 129.

4. *Letter to Members*, vol. 2, no. 6, February 1972, p. 10.

5. D. S. Greenberg, "The National Academy of Sciences: Profile of an Institution (III)," *Science*, vol. 156 (April 28, 1967), p. 492. Copyright © 1967 by the American Association for the Advancement of Science.

6. National Academy of Sciences, *Letter to Members*, vol. 3, no. 6, May 1973, p. 8.

7. National Academy of Sciences, *The Science Committee*, Washington, D.C., 1972, Appendix E, pp. 66–67.

8. Claude E. Barfield, "Science Report/Attack on Federal Radiation Standards Threatens Nuclear Development Program," *National Journal*, November 14, 1970, p. 2486.

9. National Academy of Sciences–National Research Council, *The Effects on Populations of Exposure to Low Levels of Ionizing Radiation*, Washington, D.C., 1972, pp. 7, 6.

10. National Academy of Sciences, *Semiannual Report by the Committee on Motor Vehicle Emissions*, Washington, D.C., January 1, 1972, pp. 4–5.

11. National Academy of Sciences, *Technology: Processes of Assessment and Choice*, published as committee print for House Committee on Science and Astronautics, U.S. Government Printing Office, Washington, D.C., July 1969, p. 66.

12. U.S. Senate, Committee on Rules and Administration, *Office of Technology Assessment for the Congress*, Hearings before the Subcommit-

tee on Computer Services on S. 2302 and H.R. 10243, 92nd Cong., 2nd Sess., March 2, 1972, pp. 85–86.

13. Calvin H. Cobb, Jr. (of Steptoe & Johnson, the Academy's law firm), to Ronald L. Plesser (a Ralph Nader associate specializing in freedom of information law), September 24, 1973.

14. U.S. Congress, House *Congressional Record*, 92nd Cong., 2nd Sess., vol. 118, pt. 24, September 18, 1972, p. 30954.

15. *Congressional Record*, vol. 118, pt. 24, September 20, 1972, p. 31421.

16. Federal Advisory Committee Act, Section 10(e) and (f). Reprinted in *Congressional Record*, vol. 118, pt. 24, September 18, 1972, p. 30953.

17. *Ibid.*, Section 10(a), (b), (c), p. 30953.

18. Ronald L. Plesser to Calvin H. Cobb, Jr., June 19, 1973.

19. Calvin H. Cobb, Jr., to Ronald L. Plesser, September 24, 1973.

20. Milton Katz, address before the annual banquet of the National Academy of Sciences, Washington, D.C., April 25, 1972.

21. *Science & Government Report*, vol. 3, no. 16 (September 15, 1973), p. 7.

22. Philip Handler, President's Report to the Annual Meeting, April 25, 1972. Reprinted in *Letter to Members*, vol. 2, no. 8 (May 1972), pp. 7–8.

23. *Letter to Members*, vol. 1, no. 8 (June 1971), pp. 2–3.

24. Frank B. Jewett, "The Academy—Its Charter, Its Functions and Relations to Government," read in closed session before the Academy on November 17, 1947, published in *Proceedings of the National Academy of Sciences*, vol. 48, no. 4 (April 15, 1962), p. 482.

APPENDIX A

AN ACT TO INCORPORATE THE NATIONAL ACADEMY OF SCIENCES

Be it enacted by the Senate and House of Representatives of the United States of America in Congress assembled, That Louis Agassiz, Massachusetts; J. H. Alexander, Maryland; S. Alexander, New Jersey; A. D. Bache, at large; F. B. Barnard,[1] at large; J. G. Barnard, United States Army, Massachusetts; W. H. C. Bartlett, United States Military Academy, Missouri; U. A. Boyden,[2] Massachusetts; Alexis Caswell, Rhode Island; William Chauvenet, Missouri; J. H. C. Coffin, United States Naval Academy, Maine; J. A. Dahlgren,[2] United States Navy, Pennsylvania; J. D. Dana, Connecticut; Charles H. Davis, United States Navy, Massachusetts; George Engelmann, Saint Louis, Missouri; J. F. Frazer, Pennsylvania; Wolcott Gibbs, New York; J. M. Giles,[3] United States Navy, District of Columbia; A. A. Gould, Massachusetts; B. A. Gould, Massachusetts; Asa Gray, Massachusetts; A. Guyot, New Jersey; James Hall, New York; Joseph Henry, at large; J. E. Hilgard, at large, Illinois; Edward Hitchcock, Massachusetts; J. S. Hubbard, United States Naval Observatory, Connecticut; A. A. Humphreys, United States Army, Pennsylvania; J. L. Le Conte, United States Army, Pennsylvania; J. Leidy, Pennsylvania; J. P. Lesley, Pennsylvania; M. F. Longstreth, Pennsylvania; D. H. Mahan, United States Military Academy, Virginia; J. S. Newberry, Ohio; H. A. Newton, Connecticut; Benjamin Peirce, Massachusetts; John Rodgers, United States Navy, Indiana; Fairman Rogers, Pennsylvania; R. E. Rogers, Pennsylvania; W. B. Rogers, Massachusetts; L. M. Rutherfurd, New York; Joseph Saxton, at large; Benjamin Silliman, Connecticut; Benjamin Silliman, Junior, Connecticut; Theodore Strong, New Jersey; John Torrey, New York; J. G. Totten, United States Army, Connecticut; Joseph Winlock, United States Nautical Almanac, Kentucky; Jeffries Wyman, Massachusetts; J. D. Whitney, California; their associates and successors duly chosen, are hereby incorporated, constituted, and declared to be a body corporate, by the name of the National Academy of Sciences.

SEC. 2. *And be it further enacted,* That the National Academy of Sciences shall consist of not more than fifty ordinary members, and the said corporation hereby constituted shall have power to make its own organization, including its constitution, bylaws, and rules and regulations; to fill all vacancies created by death, resignation, or otherwise; to provide for the election of foreign and domestic members, the division into classes, and all other matters needful or usual in such institutions and to report the same to Congress.

SEC. 3. *And be it further enacted,* That the National Academy of Sciences shall hold an annual meeting at such place in the United States as may be designated, and the Academy shall, whenever called upon by any depart-

[1]The correct name of this charter member was F. A. P. Barnard.
[2]Declined.
[3]The correct name of this charter member was J. M. Gilliss.

ment of the Government, investigate, examine, experiment, and report upon any subject of science or art, the actual expense of such investigations, examinations, experiments, and reports to be paid from appropriations which may be made for the purpose, but the Academy shall receive no compensation whatever for any services to the Government of the United States.

GALUSHA A. GROW,
Speaker of the House of Representatives.
SOLOMON FOOT,
President of the Senate pro tempore.

Approved, March 3, 1863.
ABRAHAM LINCOLN, *President.*

AMENDMENTS

AN ACT To amend the act to incorporate the National Academy of Sciences

Be it enacted by the Senate and House of Representatives of the United States of America in Congress assembled, That the act to incorporate the National Academy of Sciences, approved March third, eighteen hundred and sixty-three, be, and the same is hereby, so amended as to remove the limitation of the number of ordinary members of said Academy as provided in said act.

Approved, July 14, 1870.

AN ACT To authorize the National Academy of Sciences to receive and hold trust funds for the promotion of science, and for other purposes

Be it enacted by the Senate and House of Representatives of the United States of America in Congress assembled, That the National Academy of Sciences, incorporated by the act of Congress approved March third, eighteen hundred and sixty-three, and its several supplements be and the same is hereby, authorized and empowered to receive bequests and donations and hold the same in trust, to be applied by the said Academy in aid of scientific investigations and according to the will of the donors.

Approved, June 20, 1884.

AN ACT To amend the act authorizing the National Academy of Sciences to receive and hold trust funds for the promotion of science, and for other purposes

Be it enacted by the Senate and House of Representatives of the United States of America in Congress assembled, That the act to authorize the National Academy of Sciences to receive and hold trust for the promotion of science, and for other purposes, approved June twentieth, eighteen hundred and eighty-four, be, and the same is hereby, amended to read as follows:

"That the National Academy of Sciences, incorporated by the act of Congress approved March third, eighteen hundred and sixty-three, be, and the same is hereby, authorized and empowered to receive by devise, bequest, donation, or otherwise, either real or personal property, and to hold the same absolutely or in trust, and to invest, reinvest, and manage the same in accordance with the provisions of its constitution, and to apply said property and the income arising therefrom to the objects of its creation and according to the instructions of the donors: *Provided, however,*

That the Congress may at any time limit the amount of real estate which may be acquired and the length of time the same may be held by said National Academy of Sciences."

SEC. 2. That the right to alter, amend, or repeal this act is hereby expressly reserved.

Approved, May 27, 1914.

APPENDIX B

EXECUTIVE ORDERS GOVERNING NATIONAL RESEARCH COUNCIL

1. EXECUTIVE ORDER ISSUED BY THE PRESIDENT OF THE UNITED STATES, MAY 11, 1918

The National Research Council was organized in 1916 at the request of the President by the National Academy of Sciences, under its congressional charter, as a measure of national preparedness. The work accomplished by the Council in organizing research and in securing cooperation of military and civilian agencies in the solution of military problems demonstrates its capacity for larger service. The National Academy of Sciences is therefore requested to perpetuate the National Research Council, the duties of which shall be as follows:

1. In general, to stimulate research in the mathematical, physical, and biological sciences, and in the application of these sciences to engineering, agriculture, medicine, and other useful arts, with the object of increasing knowledge, of strengthening the national defense, and of contributing in other ways to the public welfare.
2. To survey the larger possibilities of science, to formulate comprehensive projects of research, and to develop effective means of utilizing the scientific and technical resources of the country for dealing with these projects.
3. To promote cooperation in research, at home and abroad, in order to secure concentration of effort, minimize duplication, and stimulate progress; but in all cooperative undertakings to give encouragement to individual initiative as fundamentally important to the advancement of science.
4. To serve as a means of bringing American and foreign investigators into active cooperation with the scientific and technical services of the War and Navy Departments and with those of the civil branches of the Government.
5. To direct the attention of scientific and technical investigators to the present importance of military and industrial problems in connection with the war, and to aid in the solution of these problems by organizing specific researches.
6. To gather and collate scientific and technical information, at home and abroad, in cooperation with governmental and other agencies, and to render such information available to duly accredited persons.

Effective prosecution of the Council's work requires the cordial collaboration of the scientific and technical branches of the Government, both military and civil. To this end representatives of the Government, upon the nomination of the National Academy of Sciences, will be designated by the President as members of the Council, as heretofore, and the heads of the departments immediately concerned will continue to cooperate in every way that may be required.

WOODROW WILSON.

THE WHITE HOUSE, *May 11, 1918.*

(No. 2859)

NATIONAL RESEARCH COUNCIL

EXECUTIVE ORDER ISSUED BY THE PRESIDENT OF THE UNITED STATES, MAY 10, 1956

Executive Order No. 2859 of May 11, 1918, relating to the National Research Council is hereby amended to read as follows:

"NATIONAL RESEARCH COUNCIL OF THE NATIONAL ACADEMY OF SCIENCES

"Whereas the National Research Council (hereinafter referred to as the Council) was organized in 1916 at the request of the President by the National Academy of Sciences, under its congressional charter, as a measure of national preparedness; and

"Whereas in recognition of the work accomplished by the National Academy of Sciences through the Council in organizing research, in furthering science, and in securing cooperation of government and non-government agencies in the solution of their problems, the Council has been perpetuated by the Academy as requested by the President in Executive Order No. 2859 of May 11, 1918; and

"Whereas the effective prosecution of the Council's work requires the close cooperation of the scientific and technical branches of the Government, both military and civil, and makes representation of the Government on the Council desirable:

"Now, Therefore, by virtue of the authority vested in me as President of the United States, it is ordered as follows:

"1. The functions of the Council shall be as follows:

"(a) In general, to stimulate research in the mathematical, physical, and biological sciences, and in the application of these sciences to engineering, agriculture, medicine, and other useful arts, with the object of increasing knowledge, of strengthening the national defense, and of contributing in other ways to the public welfare.

"(b) To survey the broad possibilities of science, to formulate comprehensive projects of research, and to develop effective means of utilizing the scientific and technical resources of the country for dealing with such projects.

"(c) To promote cooperation in research, at home and abroad, in order to secure concentration of effort, minimize duplication, and stimulate progress; but in all cooperative undertakings to give encouragement to individual initiative, as fundamentally important to the advancement of science.

"(d) To serve as a means of bringing American and foreign investigators into active cooperation with the scientific and technical services of the Department of Defense and of the civil branches of the Government.

"(e) To direct the attention of scientific and technical investigators to the importance of military and industrial problems in connection with national defense, and to aid in the solution of these problems by organizing specific researches.

"(f) To gather and collate scientific and technical information, at home and abroad, in cooperation with government and other agencies, and to render such information available to duly accredited persons.

"2. The Government shall be represented on the Council by members who are officers or employees of specified departments and agencies of the executive branch of the Government. The National Academy of Sciences shall specify, from time to time, the departments and agencies from which Government members shall be designated, and shall determine, from time to time, the number of Government members who shall be designated from each such department and agency. The head of each such specified department or agency shall designate the officers and employees from his department or agency, in such numbers as the National Academy of Sciences shall determine, who shall be members of the Council, but shall designate only those persons who are acceptable to the Academy."

This order shall not be construed as terminating the tenure of any person who has heretofore been designated as a member of the Council.

(Signed) DWIGHT D. EISENHOWER.

THE WHITE HOUSE, *May 10, 1956.*

(No. 10668)

INDEX